Out of the Labyrinth

# Out of the Labyrinth:
## Setting Mathematics Free

ROBERT KAPLAN AND ELLEN KAPLAN

Illustrations by Ellen Kaplan

OXFORD
UNIVERSITY PRESS

2007

# OXFORD
UNIVERSITY PRESS

Oxford University Press, Inc., publishes works that further
Oxford University's objective of excellence
in research, scholarship, and education.

Oxford   New York
Auckland   Cape Town   Dar es Salaam   Hong Kong   Karachi
Kuala Lumpur   Madrid   Melbourne   Mexico City   Nairobi
New Delhi   Shanghai   Taipei   Toronto

With offices in
Argentina   Austria   Brazil   Chile   Czech Republic   France   Greece
Guatemala   Hungary   Italy   Japan   Poland   Portugal   Singapore
South Korea   Switzerland   Thailand   Turkey   Ukraine   Vietnam

Published by Oxford University Press, Inc.
198 Madison Avenue, New York, NY 10016
www.oup.com

Oxford is a registered trademark of Oxford University Press

Library of Congress Cataloging-in-Publication Data
Kaplan, Robert, 1933–
Out of the labyrinth : setting mathematics free / Robert and Ellen Kaplan.
p. cm.
Includes index.
ISBN-13: 978-0-19-514744-5
1. Mathematics—Study and teaching—United States.
I. Kaplan, Ellen, 1936–
II. Title.
QA13.K37 2006
510.71—dc20
2006020661

Also by Robert Kaplan
*The Nothing That Is: A Natural History of Zero*
Also by Robert & Ellen Kaplan
*The Art of the Infinite: The Pleasures of Mathematics*

Photograph, page 6: Copyright © Eve Arnold/Magnum Photos
Page 58: "Mind," by Richard Wilbur. Reprinted from *New and Collected Poetry*,
by Richard Wilbur (1983), with the permission of Harcourt, Brace & Jovanovich.

1 3 5 7 9 8 6 4 2
Printed in the United States of America
on acid-free paper

*To all our students and
colleagues of The Math Circle*

# Contents

Theorem of Lagrange
Any quadratic surd has a eventually periodic

# Contents

# Acknowledgments

Our thanks to our colleagues: Tomás Guillermo, Mira Bernstein, and Jim Tanton, who at different times ran The Math Circle with us; to Barry Mazur and Andrei Zelevinsky, guiding spirits; to Ed Friedman, Nina Goldmakher, Danny Goroff, Patrick and Hue Tam Tai, and Marleen Winer for all their help; to the Clay Foundation, Andrei Shleifer, Jean de Valpine, and the Herbert O. Wolfe Foundation for their generosity; and to all those who have taught and worked with us: Ivan Corwin, Jim Corwin, Aaron Dinkin, Daphne Dor-Ner, Andy Engelward, Ken Fan, Leo Goldmakher, Thomas Hull, Sonal Jain, Tim Johnson, Julia Hanover, Kate Hendrix, Jesse Kass, Abhinav Kumar, Sam Lichtenstein, Jacob Lurie, Erick Matsen, Irene Minder, Vivek Mohta, Ronen Mukamel, Gautam Mukunda, Charles Nathanson, Judy Obermayer, Adi Ofer, David Pollack, Rob Pollack, Katie Rae, Benjamin Rapoport, Sarah Raynor, Yakir Reshef, Gordon Ritter, Noah Rosenblum, Jim Schwartz, Amanda Serenevy, Frederick Simons, Nadine Solomon, Angela Vierling-Claassen, Jared Wunsch, and Dan Zaharopol.

We're very lucky to have, in Peter Ginna and Michael Penn, those rarest of editors, whose thoughts are in harmony with their authors', and who help them bring their voices into the public conversation with care, humor, and grace.

Doubly lucky too in Eric Simonoff and Cullen Stanley, of Janklow & Nesbit, who have made smooth all the rough places between idea and realization.

The writing of this book was in part supported by a grant from the Cookie Jar Foundation.

Out of the Labyrinth

# one

# A Glimpse Inside

When I walked into the Harvard classroom, about eight students were already there, as I'd expected. But people passing by were surprised to see them—not because there was anything unusual in early arrivals at the beginning of a semester but because of what these students looked like: sitting in their tablet armchairs, their little legs sticking straight out in front of them. They were all about five years old.

This was the first class for these new members of The Math Circle. The parents, ranged at the back, may have been nervous but their children weren't, because it was simply the next thing to happen in a still unpredictable life.

"Hello," I said. "Are there numbers between numbers?"

"I don't know what you're talking about," said Dora, in the front row.

"Oh—well, you're right—I'm not too sure myself." I drew a long line on the blackboard and put a 0 at one end, a 1 near the other. "Is there anything in there?"

```
 _____
 0                            1
```

Sam jumped up and down. "No, there's nothing there at all," he exclaimed, "except of course for one half." This wasn't as surprising a remark as you might think, since Sam had just turned an important five and a half.

"Right," I said, and made a mark very close to the 0 and carefully labeled it 1/2.

```
 _____
 0    ½                       1
```

1

"It doesn't go there!" said Sonya, sitting next to Dora.

"Really? Why not?"

"It goes in the middle."

"Why?"

"Because that's what one half *means!*"

I erased my mark and, with a show of reluctance, moved 1/2 to the middle, acting now as no more than Sonya's secretary.

"Well," I said, "is there anything else in there?"

Silence . . . five seconds of silence, which in a classroom can seem like five minutes. Ten seconds. Tom, at the back, got up and began to put on his jacket, because clearly the course was over: we'd found all there was to find between 0 and 1.

"I guess it's just a desert," I said at last, "with maybe a camel or a palm tree or two," and began sketching in a palm tree on the number line.

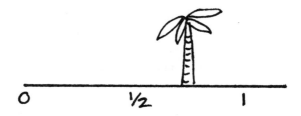

"Now that's ridiculous!" said Dora, "there can't be any palm trees there!"

"Why not?"

"Because it isn't that sort of thing!" This is an insight many a philosopher has struggled over.

Obediently I began erasing my palm tree, when I felt a hand grabbing the chalk from mine. It was Eric, who hadn't said anything up to this point. He started making marks all over the number line—most of them between 0 and 1 but a fair number beyond each.

"There are kazillions of numbers in here!" he shouted.

"Is kazillion a number?" Sonya asked—and we were fully launched.

In the course of the semester's ten one-hour classes, these children invented fractions (and their own notation for them), figured out how to compare them—as well as adding, subtracting, multiplying, and dividing them; and with a few leading questions from me, turned them into decimals. In the next to last class they discovered that the decimal form of a fraction always repeats—and in the last class of all, came up with a decimal that had no fractional equivalent, since the way they made it guaranteed that it wouldn't repeat (0.12345678910111213 . . . ).

The conversations involved everyone, even a parent or two, who had to be restrained. Boasting and put-downs were quickly turned aside (this wasn't that sort of thing), and an intense and delightful back-and-forth took their place.

A sample: after they had found all sorts of fractions in the second class (most but not all with numerator 1), and located them (often incorrectly) on the number line where they thought they should be, I began the third class by saying I was troubled by having 1/3 to the left of 1/4 (in that part of the desert between 0 and 1/2), as they had wanted.

"You may be right," I said, "but I need convincing." A long debate followed, which divided the class into those who would (for no very good reason other than reading my doubt) put 1/4 before 1/3, and those who argued for leaving things as they stood, on the grounds that since 4 was greater than 3, 1/4 had to be greater than 1/3.

At the start of the next class, Anne put up her hand. Would I please draw two pies on the board? I groaned inwardly: she must have had a conversation with her parents during the week; but I drew two large, equal circles for her.

"Now make the Mercedes sign in one in one and a cross in the other," she said. I did. We contemplated the drawings—the message seemed clear. But Anne wasn't finished.

"There, you see? If it's a pie you like, a third is bigger than a fourth! But if it's rhubarb, then a fourth is more than a third!" This wasn't an argument I had encountered before.

You probably think this was a class of young geniuses, hand-picked for special training. In fact it was like any other Math Circle class: we take whoever comes—math lovers and math loathers; we don't advertise but rely on word of mouth. True, this is in the Boston area, with its high percentage of academic and professional families, but we've had the same sort of classes with inner-city kids, suburban teenagers, college students, and retirees—from coast to coast, in London and Edinburgh, and (in German!) in Zurich and Berlin. The human potential for devising math, with pleasure, is as great as it is for creative play with one's native language: because mathematics *is* our other native language.

We came up with the idea of having such sessions one warm September evening in 1994, while having dinner with our friend Tomás. The academic year was beginning, and what had long been bothering each of us inevitably came up: how could the teaching of math be in such bad shape—especially in a city known for mathematical research? All that our students, whether college freshmen or ninth graders, knew about math was that they had to take it, and that they hated it.

We had chicken with cream and complaints, but the chocolate mousse that followed was too light for such gloom, and one of us said: "Why don't we simply start some classes ourselves?"

Tomás thought he could rent the basement of a nearby church on Saturday mornings. We wrote up a list of friends with children and telephoned around: "We're going to have some informal conversations about math this Saturday morning. Would your children be interested?"

One of those we'd spoken to had heard of Nina Goldmakher in Brookline, who ran a Russian literary circle for children, and called her up. She was wonderfully enthusiastic and sent all of her students to us. We began that first morning with twenty-nine teenagers.

Word spread. We outgrew the church basement by the end of the semester, and Professor Andrei Zelevinsky, at Northeastern University, offered us free rooms there. Parents of younger and younger students called us up; Professor Daniel Goroff at Harvard found rooms for them on weekdays. To keep classes small, we trained graduate students as teachers. We incorporated as a nonprofit in 1996, still rely only on word of mouth, averaging 120 students each semester, meeting in twelve different sections. We've kept the style informal, the bureaucracy minimal—and couldn't have more fun.

What has stood in the way of enjoying and mastering math as one does music, how we have removed these barriers in The Math Circle, and how such circles for young and old may now spread, is the subject of this book.

# two

# Cod Liver Oil

A Puritan twist in our nature makes us think that anything good for us must be twice as good if it's hard to swallow. Learning Greek and Latin used to play the role of character builder, since they were considered to be as exhausting and unrewarding as digging a trench in the morning and filling it up in the afternoon. It was what made a man, or a woman—or more likely a robot—of you. Now math serves that purpose in many schools: your task is to try to follow rules that make sense, perhaps, to some higher beings; and in the end to accept your failure with humbled pride. As you limp off with your aching mind and bruised soul, you know that nothing in later life will ever be as difficult.

> Histories make men wise; poets, witty; the mathematics, subtile.
> —Francis Bacon

What a perverse fate for one of our kind's greatest triumphs! Think how absurd it would be were music treated this way (for math and music are both excursions into sensuous structure): suffer through playing your scales, and when you're an adult you'll never have to listen to music again. And this is mathematics we're talking about, the language in which, Galileo said, the Book of the World is written. This is mathematics, which reaches down to our deepest intuitions and outward toward the nature of the universe—mathematics, which explains the atoms as well as the stars in their courses, and lets us see into the ways that rivers and arteries branch. For mathematics itself is the study of connections: how things ideally must and, in fact, do sort together—beyond, around, and within us. It doesn't just help us to balance our checkbooks; it leads us to see the balances hidden in the tumble of events, and the shapes of those quiet symmetries behind the random clatter of things. At the same time, we come to savor it, like music, wholly for itself. Applied or pure, mathematics gives whoever enjoys it a matchless self-confidence, along with a sense of partaking in truths that follow neither from persuasion nor faith but stand foursquare on their own. This is why it appeals to what we will come back to again and again: our *architectural instinct*—as deep in us as any of our urges.

Look around you: trees grow in countless shapes, birds in their infinite variety dart among them, we stroll in their shade, intent on divergent ends—yet we and the birds and the trees can only live and move because our collection of parts, and the forces acting on them, are in an ever-changing equilibrium, like a Calder mobile. The parts and the forces alike anatomize down to triangles, and these balance—singly and collectively—only because, no matter how different one triangle is from another, each has a center of gravity: the single point where its three medians meet. Nothing seems to demand that there always *be* this single point: why couldn't two of the medians meet at one place and the third meet each of them elsewhere? Why should such a coincidence happen in any triangle, much less all? And yet we can prove, with a logical argument impervious to rhetoric, that this source of our animation must lie in every triangle that ever was, or is, or could be. The argument is of our devising, but it darts up all at once, away from our here and now and our strolling personae, like a bird, high above the trees, and even beyond the birds, to a timeless understanding of the hang of things.

> It was a felicitous expression of Goethe's to call a noble cathedral "frozen music", but it might even better be called "petrified mathematics."—J. W. A. Young

This is the serene countenance of mathematics. How did so shining a beauty turn into such a wicked witch? Very young children are obsessed with numbers, counting everything in sight and playing counting-out games and fooling around with numbers and their names and their symbols the way they fool around with words in that exuberant gaiety that can blossom to meaning-bearing metaphor.

Then comes school.

By the time children are twelve the fun has all been leached out, and dread competes with boredom for its place. They begin to feel that math is in league with their enemies to make fools of them.

We teachers are largely to blame. Many of us are ourselves afraid of math. So many elementary school teachers love reading or writing or history or civics or ecology—math comes last on the list. And why shouldn't it? The traumatized child is father to the traumatized man. Few teachers have a secure sense of how what they teach dovetails into the whole (or even that there *is* a whole). They aren't sure what justifies the stupid rules you have to memorize ("'My Dear Aunt Sally': Multiply first, then Divide, then Add, last, Subtract").* They become the hated tax collectors on their students' fund of good will, sent out by a remote authority that publishes Edicts of Right Answers. Of course they end up doing their best teaching not of math but of their *fear* of it. How could they not conspire with their victims to get through the hours and days and weeks and months as quickly and unquestioningly as possible?

We've described the worst of these teachers. What about the better ones? They want their students to understand how to add 1/3 and 1/5, and so show them that 1/2 and 1/3 can be rewritten as 3/6 and 2/6, which add up to 5/6. And 1/4 + 1/5 is the same as 5/20 + 4/20, making 9/20 . . . so look, 1/3 + 1/5 turns into what? 5/15 + 3/15 . . . and that's . . . yes . . . 8/15.

Good; but do two examples prove a truth? Do twenty, or even two hundred? Are teachers going to derive the notion of least common multiple instead from the axioms for arithmetic (or, shades of the 1970s, from set theory!), and send their ten-year-olds screaming into the street? Are they going to sidestep the whole issue by turning fractions into decimals and letting fingers solicit the answer from the oracle of calculators, so that we all become the slaves of our mechanical servants?

Even teachers with the best intentions in the world find themselves compressed under the moving hand of the clock, which leaves neither leisure to explore nor the vital pause for doubting, and certainly no room to follow conjectures wherever they may lead. Everyone (except mathematicians) knows there is only one right answer. Everyone (except mathematicians) knows your thinking has to start salivating when the bell

---

*We've encountered a curious, aggressive response to this fear of math on the part of a few primary and secondary school teachers. "I know all about math," one of them told us, "and don't need that fancy stuff they do in college. This is the real thing, which those people in their ivory towers haven't a clue about." That is, she knew what $7 \times 8$ was, without having to have it dressed up in the language of sets or of axiom systems. If we take what she said seriously, and not just as living in denial, it might well follow from her recognizing that math is as much our native language as is English—for which you don't need grammar books in order to speak correctly. But of course her view stops short at the surface of math's usage—the "number facts", and never enters the depths where the reasons for them swim.

rings, and run panting until the next bell, and then stop, so that what were games become contests, and delight is turned into anxiety, and getting that right answer becomes the goal, instead of understanding.

Behind these many ways of blunting thought stands a problem peculiar to mathematics. Look back on your own career: you may have had only one good teacher of literature or history or language, preceded and followed by worse—but that one bright spot lit up the whole field and preserved it for you, so that you could take pleasure in reading or writing on your own. But learning mathematics is linked and linear: once you trip up, all the good teaching in the world is not likely to set you again on its path. Fall from a ledge and the odds are slim that you'll climb back up to and past it.

Here are places at each of which many students have lost their footing. Each of us remembers where it happened to us:

> place value
> negative numbers
> long multiplication, and worse, long division
> adding fractions
> letters for numbers
> x, the unknown
> x, the variable
> rigorous proof
> imaginary numbers
> limits and calculus
> non-Euclidean geometry
> topology
> category theory

In everything else we do there are degrees and stages of mastery, so that we can not only see where we're going and how far we've come but can in fact enjoy ourselves at each level, with that fine capacity of ours for imagining the sandlot to be Yankee Stadium. The novice guitarist relishes the music he makes, strumming one or two chords behind the vocal and losing all sense of self in the performance: the amateur by definition *loves,* and love knows no hierarchy. Even professional athletes will remember with more pleasure a sunlit game when they were ten than a triumph at the peak of their careers. Only in mathematics does mastery of one level seem to count for nothing when entering the next. How did learning place value, and memorizing your times tables, prepare you for figuring out how to add 1/3 + 1/5? Back to square one! Mathematics makes mountains out of such molehills, and a Sisyphus out of each of its climbers.

Why should this be? The reasons go down to the very roots of the art. People must have struggled for a long time with adding fractions before someone thought it through and realized that you can only add together things of the same sort (3 apples and 5 oranges are either just 3 apples and 5 oranges—*or* 8 pieces of fruit). You need to turn thirds and fifths into the same kind of fraction. Then comes the second, tactical, insight that not eighths (3 + 5) but fifteenths (3 × 5) were this kind (and why is that? Has it something to do with *dividing* a pie into pieces?). To teach it now as if it were A Rule, or (even more intimidating), The Law, is to pretend that what took years of experiment and ingenuity is as obvious as your nose. And then, because you never really had a chance to understand what was going on ("A negative times a negative is positive, because that's the way it is!"), whenever you need this rule again it will come as just that—an arbitrary fiat, enforced by Them. And so the whole integrity of mathematics is compromised. The only reasonable conclusion for a struggling student to draw from such pretense is that he is irremediably stupid, or that Mathematics works in mysterious ways, its wonders to perform.

Kant said that mathematics is synthetic a priori: synthetic because we invent it, and a priori—prior to any experience—because we then see our inventions as discoveries. It is very hard to believe that the way to add fractions wasn't already "out there"—wherever "there" is: independent of us and our hit-or-miss contrivings. Out there with the fractions—but of course they too were invented by us. We certainly are still very far from understanding the relation between thought, mathematics, and the world; but our ignorance is no excuse for pretending to others that what took effort (and perhaps well-prepared luck) to grasp should now be obvious to all. A teacher we knew told a student to take his feet off the desk. "I've been telling people for twenty years to take their feet off the desk!" Yes, but it was *this* student's first day in school.

We fall so easily into taking new mathematical insights as eternal knowledge for several reasons. Once you get the hang of a tool you want to use it and, intent on the application, tend to forget or ignore the time and thought it took to get that hang. If you're in the Tour de France, having learned to balance on a bike belongs to a prior existence. It is an understandable failure of imagination, then, coupled with impatience, that leads a teacher into thinking that what is obvious to him must be obvious to his student, so let's get on with it. It is likely, too, that many a teacher of mathematics hasn't realized that math has a history (a fault of the way we train our teachers, as if standing on the shoulders of giants meant we had no need to look down), and so a teacher, who is supposed to develop our powers of inquiry, becomes instead a messenger of Received Truth.

Mathematical shorthand plays a role here too. It is so easy and quick to write $1/3 + 1/5 = 5/15 + 3/15 = 8/15$ that we're swept past the thinking by the notation. If you can say it that quickly, there must not be that much to it. Impatience, slick symbols, and lack of imagination are external forces etching the ledges of mathematics. But these ledges are also intrinsic to the singular material of mathematics, which is *structure*. What can you get your hands on, what can you see? You may see three apples and five apples, and by an act of imagination (for regrouping is nothing less), realize that you have eight apples—but you never see "3" and "5" bare. Not even Euclid did. To recognize that behind the apples and pears, $3 + 5 = 8$, is an act of abstraction unique, perhaps, to our species. How can we see $1/3$ and $1/5$? Taking two pies and cutting one up into thirds and the other in fifths may help; abstracting the pies to circles on the board and doing your dissections there may help yet more; but it isn't even the symbols "1/3" and "1/5" we're talking about; it's what they stand for, which you can't picture. Even the mind-boggling abstraction:

$$\frac{a}{b} + \frac{c}{d} = \frac{ad + bc}{bd}$$

*points to* the immaterial relation we have in mind, which as teachers we would like another to have in mind too, and so grasp in all its generality. We have to recognize that this is an undertaking like no other (save perhaps music). The handholds seem to grow fewer the higher you climb.

Mathematics is all ledges. You no sooner acclimate yourself to breathing the thin air at this new height than the way opens up to one still higher (because math is freely—you might almost think arbitrarily—invented). No sooner do you feel comfortable at turning your new insight into the condensed form of symbols that can be used mechanically (because math is a priori: it underlies our thought), than you have to start thinking from scratch again about what some combination of these symbols suggests. $1/3 + x = 1/5$: what am I supposed to do with that? This $x$ isn't like the $a$, $b$, $c$, and $d$ in $a/b + c/d$, which were placeholders for any numbers (or almost any numbers): it is an unknown let loose like a wolf in the sheepfold of fixed quantities. Still, it is an unknown, which, if I can just contrive a way, I can make known as a fixed quantity too. But what about

$$y = \frac{1}{3}x + \frac{1}{5}$$

what is $x$, what are $x$ and $y$, now? Not fixed but variable quantities (if such a term makes any sense). Ten-year-olds the world round beg their teachers: "Won't you please tell us after class what $x$ *really* is?" And everything we've mastered so far seems to count for nothing; it all slips around and the very nature of number is put in doubt. Those ledges we've already struggled up hardly needed stepping over, compared to these rising endlessly above—and they rise not continuously but by jumps: new ways of looking, new definitions, new axioms that make you brace yourself for pushing the rock once more up the mountain.

Those who love to climb mountains have a very different view of them, and it may be no accident that so many mathematicians are also mountain walkers and climbers. It isn't just the exhilaration of solving the rock face, but the fresher air along the way and the long views from the top that draw them on. "O height!" exclaimed Petrarch, the first man, they say, to climb a mountain for pleasure—and so say we in The Math Circle. We aim to take acrophobia away by having our students do the climbing however they will, with us as their Sherpas. We bring up the supplies and peg down the base camp; we point out an attractive col or a dangerous crevasse; but *they* do the exploring on a terrain we've brought them to.

> Mathematics is the science which draws necessary conclusions.
> —Benjamin Pierce

Those kids you glimpsed finding fractions between 0 and 1 will either come up with the problem of adding fractions themselves, or we will drop the question, when it seems most natural to do so, into the ongoing conversation. Since it *is* a conversation, our questions are no more intrusive than theirs: they come in sideways, not from above. So absorbed are they in their creation that they rarely ask if we know what the answer is (and if they do, we answer with a question or a suggestion that will face them around in a useful direction). Because math is a human enterprise and they are humans, their thoughts can't help but tack around the direct line. Look at a prehistoric ovoid hand-axe from Perigord and a triangular hand-axe from Tanzania: different in detail, their structure is the same because they solve, as humans would, a human problem.

Surprising and delightful things will come up. A common multiple, other than the least, of two denominators may be the first to surface, and even should the least then appear, affection for the first-born—or the delight that small children take in large numbers—may lead them to settle on the larger one.

"1/3 is 20/60 and 1/5 is 12/60, so 1/3 + 1/5 is 32/60!" exclaimed Tom. This puzzled Sonya, who had gotten $1/3 + 1/5 = 15/45 + 9/45 = 24/45$, and was sure she was right. That there were two different answers to the same question—and that both turned out to be right—was a revelation whose metaphorical value stayed with them.

They do the exploring, but each of us who leads a Math Circle course has clearly in mind from the start what the goal is; wandering around aimlessly is as boring as being marched from point to point. I wanted my young students to develop the arithmetic of fractions, to relate fractions to decimals, and at the end to invent a number that wasn't a fraction. But had they followed another line of thought, which seemed to me fruitful—from least common multiples to greatest common divisors, say, and to the Euclidean algorithm, or primes—I would have unpegged my tent and moved it to their line (and indeed I have, in other semesters when the opening question was the same, but the conversation took off entirely differently). These are decisions that have to be made more or less on the spot, and are one of the quick-reflex challenges that make Math Circle teaching exciting fun. You're in there doing math with other minds; the goal is clear but the ways to it are up for grabs and at the mercy of temperament and insight. It is just about the most exhilarating thing you can do.

Because the arena is conversation rather than competition, the students—whether they are five or fifteen or fifty—are first startled by, then quickly come to relish, other points of view. They take a collective pride in each other's insights: "How did you ever think of that!" Since the majority of suggestions that come up are faulty or imprecise or incomplete, fear and embarrassment disappear in a mutual mulling over and reshaping of questions and answers. They like the image we've passed on to them from the mathematician Barry Mazur: we are all very small mice gnawing at a very big piece of cheese; no shame, then, in having bitten into a hole, nor any need to hoard up precious crumbs.

Rhythms of different depths pulse through an hour's class and the ten weeks of a course. With the very young, attention flags after twenty minutes or so, or a threshold of frustration is reached. Time for their favorite sport: function machines. One of them announces that she has a secret rule; we put in numbers, she tells us what comes out (the Sherpa is now a blackboard amanuensis, keeping up an input-output chart underneath a baroque drawing of a function machine,

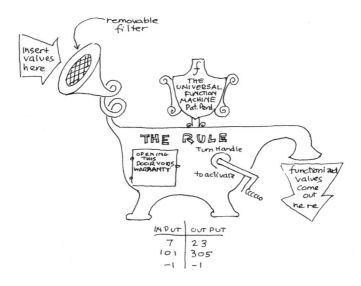

or later, letting the machine "draw itself"—off-handedly introducing them to graphing, that is—and we almost always manage to guess the rule eventually, to universal delight). Then back to where we were, with spirits refreshed.

A deeper rhythm is the alternation between intuition and proof. An insight will usually be tested by several examples and, if it looks sound, will be promoted to a conjecture. But too many likely conjectures turn out to be faulty after awhile, and the question of trying to derive our conjectures from foundations comes up. What foundations? How derive? How tell a false proof from a true one? This plays out in an unpredictable variety of ways. The point to notice is that rigor has its place: namely, where you can't do without it. The need has to arise naturally and has to be intrinsic to the enterprise, as if it were a biological necessity. The younger students are often buoyed by waves of enthusiasm for a conjecture that we'll conclude looks good, or even very good—but the shadow of that Scottish verdict, "Not Proven", is always on the wall.

Waves of enthusiasm: it's the sheer fun of it all that keeps eight-year-olds from leaving at the end of a class on a late winter afternoon, or that gets adolescents up for classes early on Sunday mornings. But before we say anything more about what works in The Math Circle and what doesn't, about how it has evolved and what directions it is likely to go in, we should really go back to those barriers between mind and mathematics: for more than the ledges within and the teaching without stand in the way.

There is the myth of talent, to begin with—and after debunking it, we'll talk about what the ingredients really are that go into cooking up a mathematician. Then there is the matter of the off-putting language of mathematics: its symbols, equations, and often glacial style. We'll talk about the way not teachers but mathematics itself can be intimidating, and then about the fact that you can't just jump into it in the middle; about the nerdish character attributed to mathematicians, which would put anyone off from wanting to keep them company; and last, the question so many ask when frustrated by a problem: Who Cares? Why should I spend a minute, much less my life, on this glass-bead game?

We'll look at some of the ways math has been taught, and what The Math Circle approach owes to them; then at how mathematicians actually do their thinking. At that point we'll be ready to look in detail at The Math Circle.

## three

# The Myth of Talent

The idea that aptitude for mathematics is rarer than
aptitude for other subjects is merely an illusion which is
caused by belated or neglected beginners.—J. F. Hebart

"I have a damned kink in my brain," said the American logician and
philosopher C. S. Peirce, and it is a commonplace of our culture that to
be a mathematician you need such a kink, damned or blessed.

Our approach to math is based on the idea that anyone can learn to
think like a mathematician and will, in the process, come to find pleasure in learning and creating math. Belief in talent is a holdover from
simpler times, when what race you belonged to determined your character and the bumps on your head revealed it. It is much more romantic
to picture geniuses, larger than life, suddenly bursting into view with
something magical, almost diabolic, about them (Liszt, Ramanujan,
Byron), than to take away the cape and the penetrating gaze and ask
what the true story is. It also absolves the rest of us from responsibility:
if a talent for math is inborn, no need to waste our time trying to develop one. Teachers are exonerated too: it is the Nature of Things that
produces a stream of mathphobic failures.

What is the true story? The masters of any craft can off-handedly do
what seems miraculous to outsiders: watch a magician shuffle a pack of
cards and divide it into eight ordered sequences. Part of a long appren-
ticeship is served in learning techniques and trying them out in varied
circumstances, until what were at first awkward gestures articulate
smoothly into a whole. To the extent that math is a craft, its skills (such
as unknotting tangled expressions and transforming the strange into
the familiar) need to be practiced. Without your knowing or willing it, a
threshold comes to be overstepped, and then you can do easily what
before you couldn't do at all (again, think of learning to ride a bicycle).
This depends not on inborn genius but on doggedness and attention.

But beyond craft, mathematics is an art, and as such calls up all our mythology about artists. The craftsman may learn fluency, but those strokes of insight that mark significant works come out of the blue. Really? Ask any artist modest enough to put mystery and the glamour of genius aside, and he will tell you they come out of the grey. For while you are at work on a problem of painting, writing, music, or math, all sorts of hooks and eyes tumble by, and connections are made only to fall apart. What was clear becomes clouded; you begin to doubt the truth you are trying to establish, and then your own powers to establish it. Good leads peter out; fresh ones discordantly clamor for your attention. You lose track of bits of the composition and forget how others worked, until it all sinks chaotically out of sight. And then the solution dawns—when your attention was on something else.

When we read poets or chemists or painters or mathematicians on how they came by their ideas, we're drawn toward agreeing that insight comes after playing intensely with the ingredients—and then leaving them to ferment in the mind's cellars, out of attention's light. Why and how this works isn't yet clear, unless it is that inattention loosens the grip of firm misconceptions. In Inca mythology the head and body of Túpac Amaru, severed by the Spaniards, are slowly working their way toward each other through the earth—and when they rejoin, his people will rise again. Something subterranean like this may be at work in the subconscious, where the fragments of thought and their reflex shapes inch closer together.*

You will likely agree that skills can be taught—but doesn't it take genius to move from good practitioner to artist? Can people be taught how to have insights? Suffice it for now to say that insights can be *prepared for,* by encouraging an imagination playful to the point of recklessness, along with a sort of experimental fervor that follows hypothesis out as if it were truth—but lets it go once revealed as error. It also helps to cultivate a healthy distrust of authority, and a restless, ranging curiosity—not so much an anarchic spirit as the flexible feel for law that tricksters (in the tradition of Odysseus) have.

The phrenological theory of bumps marking the locations of mental faculties fell away a century ago, and the notion that one is ruled by such inborn faculties is falling away now too. We are beginning to sense how adaptable the mind-brain—our organ of adaptability—really is. Given the greater sophistication of the way we understand how the mind works, why has the myth of talent persisted? When we play Mozart to newborns

---

*Another Inca theme, blended with this one, holds that the world will one day be remade as a single body, its mind in Europe, its will in North and its heart in South America. Perhaps a similar fusion of will, thought, and feeling account for insight.

in order to develop their musical sense, why do we still think a math sense is genetically determined? Certainly you can cite the Bernoulli family of mathematicians (like the musical family of Bachs), but far more often mathematicians come from parents startled by their progeny. Even among the Bernoullis, are we to deny the role that an atmosphere saturated in mathematics could have had on the rising generations?

Without any statistical evidence to back it, many still look for the mathematical rabbit to have jumped out of a genetic hat. Here are five out of probably many reasons for hugging this old husbands' and wives' tale to our bosoms.

First, we try to make sense of the extraordinary, and heeding the authority of the past, invoke the time-honored notion of a "calling", usually in a sufficiently endowed family. We shy away from asking who calls, or how many of those who have heard the call answered it well. A sacred sort of aura—almost a superstition—puts inquiry off. It may well be that when you come to feel a calling (to math or anything else), you are interpreting passively a mixture of enthusiasm and confidence, along with a barely conscious sense of the line of fall your inclinations are gliding along. A hillock climbed early, or at a significant time, may make distant mountains in that direction seem inviting rather than daunting. If you *don't* find something comes easily to you, you may believe (or society may encourage you to believe) that you have not been called to it—you are just not "meant" to be a musician or a mathematician or a poet or a parent. But you won't get any calls if you've left your phone off the hook.

Second, there is also the problem of misleading analogies. Because you do need a certain flexibility for ballet and bulk for weight lifting, fast-twitch muscles to be a successful sprinter and slow-twitch for the longer distances, we come to speak of the "muscles of the mind" and imagine that here, too, genetic determinants rule. You need only look at the variety of backgrounds from which creative mathematicians have come to let go of this hypothesis. Look also at the variety of "traits" (as they were once called) that keep mathematical creativity company to see how short this kind of argument sells the intricate reality. Some marvelous mathematicians are cultivated, some boorish; some adept in the ways of the world, others extremely innocent. You will find brilliant mathematicians reticent and aggressive, humorous and stodgy, vain and humble, lawless and law-abiding, of every political, ethnic, and social stripe—and some brilliant mathematicians who are dense outside their field. It is inertia, really, that keeps our imagination on a monorail when it approaches the farther reaches of our thinking.

Third, the romanticism we spoke of before has much to answer for. We want signs that the gods love at least some of us. We want our lives lit up by comets of genius flashing through them: we want heroes to worship, the lucky to envy, the great to glance at us where we stand by the road-side, waving. Some mathematicians connive at this romantic image too—from self-love and pride on the part of a few, and others from a fashionably detached admiration of their talents, as if they were separate beings (like the tennis players who report that "the forehand was really on today"). Most of us know few mathematicians and very seldom see them at work; a perfectly understandable aesthetic motive hides the labor preceding insight, so that only the finished theorem and its proof get published, with the scaffolding and all the fallen workmen cleared away.

Fourth, mathematics has a reputation for being an uncompromising taskmaster: the road is narrow and steep; to be wrong by a little is to be wrong altogether, and its high-altitude truths care nothing for our weaknesses. Giants they, of another species, who stand on its Everests. It does indeed take a kind of blithe fortitude to put up with years-long frustration, and such a thorough sinking of self into the subject that every smallest gain in understanding enriches the harmony of a growing whole for which we have no pronoun: neither "you" nor "it". But problems attractive enough to entice your putting in the effort develop your fortitude—who past the age of eight wants to go on beating five-year-olds at tic-tac-toe? The enticing morsels we dangle in The Math Circle are just beyond arm's reach, so that grasp lengthens along with the span of attention, and we learn to delight in contriving new ways around old obstacles. Working together prying open the lid of a treasure chest (How can you add different sorts of fractions? Are a triangle's altitudes concurrent? What is $i^i$? How could you classify all knots?) blends individual egos into the common effort and unites them at last in the insights gained.

Finally, to round out this list on a cynical note, there is certainly a living to be made out of perpetuating the myth of talent and its levels of aristocracy: "gifted", to "deeply gifted" and "profoundly gifted" (is "abysmally gifted" next?). This is a horse worth looking in the mouth. It wouldn't be hard to imagine a person setting himself up with a battery of multiple-choice tests and rewarding the parents who paid him by reporting that their child had an IQ of 183—but to keep them coming back, was also dyslexic, dyspractic, deficient in attention, or marginally autistic. Scores on intelligence tests do cover a wide range of qualities, though they are not unique determiners even of academic success: 175 is not necessarily greater than 100. Future generations will chuckle over the Binets we have in our bonnets, as we do over manuals of phrenology.

After a list such as this you may argue that we're just avoiding the blatant truth that there *are* talented young: that prodigies of calculation star the benighted past; that experiments (as with MRI) demonstrate that parts of the brain, when mechanically stimulated, produce bursts of calculating brilliance; that articles appear yearly about eleven-year-olds leaving their college classmates in the calculus dust; that there is a remarkable correlation in some children between autism and a fascination and fluency with numbers.

Rapid calculation, however, has little to do with the structural insights that make up mathematics. You need only watch a tableful of mathematicians trying to calculate the tip, to see how many are inept at numerical tasks. Some, like Gauss, are exceptionally good at them, but Gauss himself employed an even more rapid calculator named Zacharias Dase, who had no idea what mathematics was about. The great Kummer, in nineteenth-century Germany, was notoriously unable to multiply even single-digit numbers together. "Seven times nine is . . . um . . . er . . ." he said from the lectern. A mischievous student spoke up: "Sixty-one," and Kummer began to write it down. "Sixty-nine," said another. Kummer stopped. "Come come, gentlemen," he said, "it cannot be both. It must be one or the other."

> One may be a mathematician of the first rank without being able to compute. It is possible to be a great computer without having the slightest idea of mathematics.
> —Novalis

How clear is it that the symptoms we read as autistic in many mathematicians aren't effects rather than causes of their devotion to the unknown god? "Abstract thought," said Jacob Bronowski, "is the neoteny of the intellect." But you might put autism aside and ask us what we have to say about someone like the three-year-old who announced at his first Math Circle class: "I'm very numbery"—and was. Children will tell you as quickly which of their classmates is good at math as they will who is good at baseball or drawing or spelling. We don't question the presence of people in our society who are better, and in some cases very much better, than others at math. What we do ask is *how they got that way*, so that we can develop their skill in everybody.

So much of what seems innate depends on the luck of first encounters. The smiles and frowns that the world unwittingly gives to this or that little probe of ours can determine the landscape of our thought as thoroughly as a pebble diverting a rivulet will, in the end, make for a valley here and a mountain there. But behind the chance placement of the pebble lies the enormous force of gravity, just as behind our experience lies the deep structure of language: the capacity we all share for catching how the universal rules of grammar are applied in the particular language of our surroundings. That fluency with mathematics we see in some children is fluency with its *language*: another application of the same deep structure. It is unusual, of course, for this particular language to take precedence over, or even to flourish beside, our native tongue—and that's where the luck of little encounters comes in. Something most likely unnoticed in the tumble of experience catches our inner ear and makes this conversation about number, pattern, and shape as appealing as that about people and things. Most of us learn math as if it were a second language, with all the ills of translation that implies, but after years on another track, it may be that one day a shunt opens, by chance, to structure itself: an *architectural instinct,* which delights in form and whose natural expression is mathematics.*

However you decide about talent on the basis of your own experience and inclinations, you will likely agree that it is best to act *as if* it were a myth—rather in the spirit, though with the opposite polarity, of what's called Pascal's Wager (act as if God exists; for if he does you may be saved, and if he doesn't, nothing is lost). If there is no such thing as a talent for mathematics, then you do well to proceed as if everyone could be led to its delights; while if such a talent exists, your actions won't harm those who have it and may give a spectator's informed appreciation to those who haven't.

Letting the unexamined assumption of talent run inevitably leads to building tracks for it to run on, perhaps disguised by jokey names, but no one is fooled about which is for the fast and which for the slow (one school we know of labeled their tracks "Peregrines" and "Yaks"—surely not without a touch of Dickensian malice in making sure people knew their place). Prophecies by labeling, such as these, can't help but be self-fulfilling, as an experiment in a New York City public school a few years back showed: students were arbitrarily placed in two sections, but their teachers were told this was the slow group, that the fast . . . and by semester's end, so they were.

Our own experience with students certainly reinforces our conviction that "talent" is not only a myth but a pernicious one. We've had any number of students in our years of teaching who sat inert as a noble gas through class after class, and then one day blurted out a stunning insight. Where had they been until then? They may tell you (like the young Macaulay) that there just hadn't been an occasion before to speak, or they had been distracted, or hadn't caught on; those are flags variously signaling that an organic process hadn't yet passed a crucial threshold ("Why couldn't you balance on a bike before this moment?"—what would you expect a child to answer?).

We each have engraved on our hearts the names of students we had unjustly given up on. In one class of ten- to twelve-year-olds, we were coming to grips with different sizes of infinity (they were too young to

---

*Some recent experiments may suggest that linguistic and mathematical skills are distinct. These seem, however, to equate "mathematical skill" with calculation and familiarity with "number facts". In any case, we're speaking here of language in the broader sense of making and manipulating symbols for structures.

know this was hard), and flying through the central ideas of sets whose elements matched up perfectly with those of other sets, and sets where this was impossible. They made up names when they needed them for these sets and for the operations on them. Eugenia talked eagerly all the while, but always about the names and their appropriateness: this was what engaged her imagination, and the ideas that the names stood for slid harmlessly by. In the next to last class—now onto very sophisticated topics about ever larger sizes of infinity—the boy next to her was having trouble seeing why there had to be more subsets of a set than there were elements in the set itself, and Eugenia turned and explained the proof neatly to him. We all fell silent in astonishment.

"What?" said Eugenia.

"But that's perfect!" said another student.

"Well," said Eugenia, "There it is."

"But you didn't understand *anything* last week!" said the boy.

"You mean we've been doing this all along and I didn't know it?"

If you look beyond our possibly prejudiced evidence, you will find one example after another of famous mathematicians who showed no early signs of talent. The famous French mathematician Jacques Hadamard, whose accomplishments included proving the notoriously resistant Prime Number Theorem, held down the last place in his math class through seventh grade. Hermann Grassmann—one of the first to explore n-dimensional geometry—and whose hypernumbers were a daring generalization of the complex numbers, was so backward as a child that his father hoped he might at least make it as a gardener. An interest in theology took him eventually to the University of Berlin, where he never attended a single lecture in mathematics. His many important papers in mathematics and physics (not to mention his mammoth translation of the *Rig Veda*, his collection of German folksongs, publications on botany, philology, theology, and music—while raising eleven children) were posted from the provincial German town where he spent his life as a schoolteacher.

And what of that eponymous hero of superhuman genius, Albert Einstein? "I know perfectly well," he wrote, "that I myself have no special talents. It was curiosity, obsession, and sheer perseverance that brought me to my ideas. But as for any especially powerful thinking power ('cerebral muscles')—nothing like that is present, or only on a modest scale." His biographer says that Einstein was slow to talk, formulating sentences before he spoke; that he was kept home with a tutor as unready for school but flew into rages with his tutor; and that his slow wondering was the basis of his always questioning what seemed the obvious to more facile classmates.

Newton would not admit that there was any difference between him and other men, except in the possession of such habits as perseverance and vigilance. When he was asked how he made his discoveries, he answered, "by always thinking about them."—William Whewell

21

Because we discount "talent" in our classes, students confident and enthusiastic about math sit mixed in with those who have joined because, although they are unsure about math, they had heard that these classes were fun. There will always be some people in a new group who assume that the point is to be best in the class, and are quick to put themselves up and others down. We are as quick to say that the math, not our little egos, is what matters. It is remarkable how well even feisty adolescents respond to this (almost with relief), and plunge together into problems whose answer is just around the corner—though the corner seems mysteriously to keep moving on just ahead of them. They excitedly pile suggestion on suggestion and glow with delight at one another's insights. They even develop a thoughtful style of criticism:

"That's a good idea—but look, does it work in this case?"
We have even heard one rather precious six-year-old say to another: "I like your conjecture, Jeffrey, but I think I have a counterexample."

You need a level field for play to be at its best, and egos are bumpy. Self-confidence grows, we've found, when people are focused on the work at hand: pride in belonging to a group, to a species, that can crack tough problems open—pride in a world that has such secrets in it—is much more satisfying than pride in your passing triumph. This is why our response to an insight is a workmanlike "Good. Now where does that take us?" A familiar mantra is "Teach the child, not the subject", but the child is best taught while there is something external for him to think about; you don't get many new insights when contemplating your ego.

We have another reason for treating all suggestions impartially, the least likely along with the most: by letting any conjecture run—rather than stopping it with a "No, that's wrong"—we let the line of thought carry itself to a falsehood, a dead end, a more thoroughly explored uncertainty, or an onward-pointing truth; false conjectures often tell us more than true ones, narrowing down the landscape of likelihood.

Children aren't as foolish, nor adults as gullible, as the media would have us believe. Learning in order to get ahead sounds very grown up, but all the trappings of competition look pretty shoddy in the watches of the night when we ask ourselves what the point of it all is. Since as a species we survive through curiosity, a hunger to understand is at least as great a driving force in us as self-aggrandizement. Nobel Prize winners, by and large, aren't in it for the money; they are drawn on by the problem and rewarded by the admiration of their colleagues. It isn't surprising, then, that a sense of pride in the cohort develops in a Math Circle class, which leads to more pleasure in what the idea is, than in whose idea it was. This is a subtle matter, for as George Polya pointed out long ago, the secret of solving a problem in mathematics is to sink your whole personality into it—but *sinking in* is different from *slathering over*. We need our idiosyncratic approaches and points of view, but not the distracting doting on them. The myth of talent disperses the energy needed for ideas to condense.

## four

# Making Your Hats

Beauty is the first test; there is no permanent place in the world for ugly mathematics.—G. H. Hardy

A Boston matron was asked where she got her hats. "We don't get our hats," she answered haughtily, "we *have* them." Replace "hats" by "talents" and that's the myth we have just exploded. Where, then, are we supposed to get the many different hats a mathematician has to wear? We build up the skills for doing math out of traits common to us all. Let's look at three of those skills—holding on, taking apart, and putting together—with what will be a running warning that timing is all: what helps at one stage in doing math hinders at another. Keep in mind too that these capacities develop at different times and in different ways, now abetting, now interfering with, their neighbors.

## 1. Holding On

Some parts of math are slippery, some are sheer. We have to learn the mountaineer's skill of clinging on *here* while stretching out to *there*. The ingredients are already in our human nature: stubbornness and its cousin, orneriness; a threshold of frustration that can be raised, a span of attention that can be lengthened, and holding on turned inside out: putting on hold.

### Stubbornness

You have only to look at the Wright brothers' faces to know why they overcame the thousand obstacles between Dayton and Kitty Hawk: the mosquitoes, misleading tables, rain, wind, shattered spars. They hunkered down and inched forward again over much old and a little new ground.

25

In math it is often a case of the data piling up with no system to it, or a system emerging only to crumble at the next example. You come to a crossroads, and nothing in your intuition or in the landscape inclines you to follow one way rather than the other; if you try to follow both, your attention and energy drain away. In most other human callings, the terrain where you find yourself helps your moving through it, and general experience strengthens intuition—but it takes a lot of coming to know the invisible substance of math before you can make its alien-seeming abstraction a locale in your thought, and feel comfortable enough to stroll through it and begin to see that the streets have a plan and the buildings a code.

This sort of grasping is especially difficult because the frustrations in math are so definitive. They strike deep and happen often. It may be worse for novice chess players, who lose match after match before they begin to win—but at least they profit from each loss by seeing vividly that you need to look more than two (or three, or four ...) moves ahead, or that certain tactics, good in themselves, suit ill with a particular strategy. You may have to lose the first fifty times you play Go before an inkling dawns of where the sinks and sources of power are on the board. In math, however, you're beaten not by a human master but by the uncommunicative subject itself. Say you're attempting to detect a pattern among Pythagorean triples (those integers x, y, and z such that $x^2 + y^2 = z^2$). One failed attempt may follow another with you being none, or ever so little, the wiser. Just when a pattern seems to settle in (two of the three must be consecutive integers, like 3, 4, 5; 7, 24, 25; or 20, 21, 29), a counterexample spoils it (28, 45, 53). You hear the problem saying that you are too stupid to find the pattern—or that of course there is none, and you were a fool to think there was. Ghosts of pattern spread out and disperse like spume on the restless sea.

How do you develop the kind of doggedness to get through these dark moments? A good teacher can certainly help, by setting a sequence of problems that build up confidence through increasing the resistance gradually. Fellow students help too: working conversationally with others shares out the frustrations and so lessens self-doubt. Both can make the play count for more than the winning or losing. Good, we've found one Pythagorean triple (say 3, 4, 5); and another: 6, 8, 10; ah—any multiple of a triple that works will work too—that's a real gain. And there are triples that aren't multiples of the one we found (20, 21, 29 isn't a multiple of 3, 4, 5): so something has come to light and there's no danger of being bored; more revelations must lie in wait. Dig a level deeper.

The kind of tenacious oblivion needed now has a negative and a positive source. Negatively, you cast yourself in the heroic mode of opposition, invoking your favorite figure of resistance. Dancers and athletes have a well-earned sense that you can still fail despite all your totally dedicated work—and that you then rub your bruises and begin again. Positively, you focus on the particulars of the handful of dirt you've just scooped up as you tunnel doggedly forward: this pebble in it, that fragment of china. So (3, 4, 5), (20, 21, 29), (7, 24, 25) and (28, 45, 53) are Pythagorean triples—well, an odd and an even to start with, then an odd. Nothing helpful here, perhaps, but bear it in mind.

Stubbornness roots you in the terrain. We were holding a Math Circle class at Microsoft, in Seattle, for the eight- to twelve-year-old children of some researchers there. The format was a casual lunch with sandwiches. "Has anyone heard of the Pythagorean Theorem?" we asked, while munching. Many had. "OK, would someone go up and draw a right triangle on the whiteboard . . . oh, and make its legs the same length." A confident twelve-year-old boy drew a large red right triangle.

"Swell. Let's say each of its legs is a unit long." He put a "1" by each of its legs. "So how long is the hypotenuse?" A pair of twins called out, almost together: "The square root of two!"

"Right. And what is the square root of two—I mean, what number does it turn out to be?"

Fortunately there wasn't a hand calculator in sight, so they were forced to reason it out. An eight-year-old girl said 1 was too small, since $1^2$ was 1, and a ten-year-old said that 2 was too large, since $2^2$ was 4. The likely candidate 3/2 proved to be just too big, giving 9/4 when squared. We had in mind luring them into a proof, after a few frustrations, that $\sqrt{2}$ couldn't be any fraction whatever—but that wasn't the direction thought was taking this blustery March afternoon. They were going to pin down the fraction hiding behind the mask of $\sqrt{2}$, and they were going to do it before lunch was over.

Round numbers are always false.—Samuel Johnson

The twins went to the board with blue markers in their hands and, at the suggestion of the oldest girl there, patiently multiplied 14/10 by itself: 196/100, so we were definitely just about finished. While ideas raced around the table about products of two negatives and why not try decimals and whether the final answer's numerator would be even or odd, the twins quietly calculated $(142/100)^2$ and then $(141/100)^2$, with results elating or depressing, depending on your point of view. They were very fond of multiplying, and as we started to bet what size fraction would finally do it, numbers appeared out of the blue on the white sky. They actually calculated

$$(1414/1000)^2 = 1999396/1000000,$$

and—while we held our breaths—

$$(1415/1000)^2 = 2002225/1000000.$$

Now that the hunt was all but over, others joined in the calculating frenzy, often darkening the waters with conflicting results.

We asked, in a pause which owed more to catching a second wind than to fatigue, what number the numerator would have to end with so that, when squaring the fraction, the result would be all zeroes after the 2.

A voice in the wilderness: "Five in one, six in the other?"

"But they have to be the same!"

"Then it can only end in zero."

"And before the zero?"

"Before the zero—another zero—"

It wasn't so much that lights went on as that the light went *off*.

"There's nothing that works . . . "

"We've multiplied wrong . . . "

The boy who, so long ago, had drawn the triangle on the board, went silently up and erased its hypotenuse.

"You mean . . . " we began.

"This triangle has no hypotenuse," he said—and with that the class ended. (We had exactly the same experience with a group of Scottish children at the other end of the economic scale, showing that mathematics is universal even in its entrées to error: we are all novices when it comes to thinking of numbers not as objects but as sequences.)

Did the class end in failure, do you think? It seems to us rather that their *stubbornness* gave them an ineradicable sense of the terrain. The next time that hypotenuse once again sits there gleaming evilly at them—a day from now for some, a month, a year for others—they'll have a vivid context in which to approach it, and the lay of the land they traveled over together may push thought toward a next stratum down in the search for how a line can have a length that isn't rational. Imagination is born from the vivid conviction that what can't be, must be.

**Orneriness**

Isn't orneriness just another name for stubbornness? Not really. There's many a mule who won't let go of a carrot, and many a lawyer who won't let go of a line of questioning, without a touch of the prickly independence that makes a critter ornery. Their stubbornness amounts to refusing to be moved. Orneriness won the West because it does move in the direction it chooses, whatever the neighbors or the Best Authorities say.

It isn't that authority means nothing to an ornery soul. It means less than nothing: it is a negative. It stands for persuading by something other than reason, because the reasons probably aren't very good. We misread the ornery as cantankerous because of its Missourian insistence on proof.

When the third brother goes off to fight the dragon that killed the older two, he goes with fairy-tale confidence—and this, no less than convention, helps him to victory. A strong democratic sense of your being the world's equal lets you take what you will from the endless advice it offers, and then set blithely out. Someone driven by stubbornness alone pits his will against the world's, and says that he'll fight it out along this line, and this line alone, precisely because it is unpromising. Stubbornness tempered by orneriness looks for ways over, under, around, and takes what is eventually the new path as enthusiastically as the old. The chorus of people calling you a fool is so much birdsong.

Martin Hellman, one of the inventors of public key cryptography, wrote of his fellow inventor, Ralph Merkle:

> Ralph, like us, was willing to be a fool. And the way to get to the top of the heap in terms of developing original research is to be a fool, because only fools keep trying. You have idea number 1, you get excited, and it flops. Then you have idea number 2, you get excited, and it flops. Then you have idea number 99, you get excited, and it flops. Only a fool would be excited by the 100th idea, but it might take 100 ideas before one really pays off. Unless you're foolish enough to be continually excited, you won't have the motivation, you won't have the energy to carry it through. God rewards fools.

Here's an example of orneriness in action. You have an 8 × 8 square and cut it into two right-angle triangles and two trapezoids, as here:

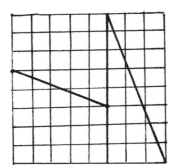

Move these four parts around and put them together again in this second configuration, whose height is 5 units and base is 13.

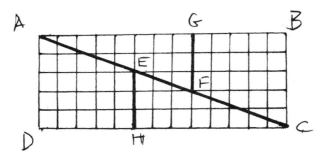

So $64 = 8 \times 8 = 5 \times 13 = 65$.

Someone sufficiently ornery might have thoughts like these: "Maybe it's true that numbers I took to be different turn out to be the same. After all, it's happened before: $e^{i\pi}$ and $-1$ are identical. But that's really a matter of a single value turning out to have two different expressions. Were there a glitch at just one point in the integers—a fold, so that 64 and 65 turned out to be the same—then in fact 1 would equal 0 and all numbers would be identical. While this is possible, it would be too un-interesting, so let's discard it on aesthetic grounds.

"Perhaps moving shapes around on the plane changes their area . . . why might area not slosh back and forth like water, and sometimes spill out? How do I know that rotation, translation, and reflection—all used here—preserve area? It's not something I've ever given much thought to. It might be interesting to imagine a geometry in which something like area—'content', in some sense—did alter with certain motions. Keep this in mind.

"Maybe cutting affects area—it certainly does in reality. Look at those shavings falling to the floor. There's the old joke test of whether or not you're a mathematician: picture a wooden cube, its top and bottom painted yellow, one pair of opposite sides painted blue, the other pair red. Now saw this cube three times, along each of its three axes. Question: did you see sawdust? If so, you're no mathematician. What a peculiar idea of how mathematicians think. Why shouldn't they, like the rest of us, be allowed to enliven conjecture by embodying it?

"Or—like sawing—maybe the thickness of the internal lines in this picture has somehow used up a square of area in the first diagram, which shows up in the second. I remember there were people in the 1920s who fitted whatever shape they wanted—Parthenon or the Mona Lisa—into rectangles that were supposed to be of the most pleasing—'Golden Mean'—proportions (like $3 \times 5, 5 \times 8, 8 \times 13$, each new number the sum of the previous two), just by drawing their enclosing rectangles with sufficiently thick pens.

"Try it conceptually, then: well, conceptually we have two congruent triangles, each with area $(3 \times 8)/2 = 12$ square units, and two trapezoids, each with bases of 3 and 5 and height 5, so each having an area of $5(3 + 5)/2 = 20$, and indeed $40 + 24 = 64$, so the first diagram is all right, and thickness of lines didn't matter. So where did the extra square unit come from in the rectangle? Gaining a 64th of the whole area seems pretty drastic (but maybe not, if distributed over the motions of four figures, with 1/256th due to each . . . )

"Well then, can you in fact dissect an $8 \times 8$ square into these two congruent triangles and two congruent trapezoids? Evidently you can. The mysterious bit comes in rearranging them, and I know there's a theorem with a longish proof about reshaping one polygon into another with the same area. So really there's something of interest here. Is this indeed a polygon we've ended with? Evidently it is. Is it a 5 × 13 rectangle? Is it a rectangle? It looks like one.

"Let's test that. Since AEFC is a straight line, ΔADC is similar to ΔEHC, and so AD/EH = DC/HC. Are these ratios the same? Well, AD/EH = 5/3 = 1.6666..., and DC/HC = 13/8 = 1.625! The triangles look similar but aren't, so AEFC wasn't a straight line after all! 'Thickness of line' wasn't all that far off the mark: the diagonal of the 'rectangle' hid a long, narrow parallelogram."

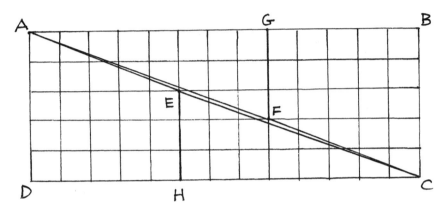

Some lessons to be learned from this. Resolve more superficial doubts before attacking profounder ones (which of course still lurk). A difference in area of a 64th is enough to fool the eye: be wary of visual proofs. The crucial proportions were 5/3 and 13/8—ratios that are indeed in the sequence that leads to the Golden Mean! Listen more closely to your intuition next time (and think more fully about why the Golden Mean should be involved in this nice deception).

## A High Threshold of Frustration

It occasionally happens, especially in younger Math Circle classes, that after working on a problem for a while one of the kids will say: "Come on, you know the answer—just tell us." There was in fact one Russian seven-year-old who never took part in the discussions but sat with his pencil poised over his pad. When we finally asked, "Andrei, what do you think?" he answered: "I'm just waiting for you to tell us, then I'll write it down."

Where does this momentary or habitual checking at the jump come from? Why have those who just now were delighting in the chase turned into Monty Python's Mr. Gumbies, crying out: "My brain hurts!" It's as if thinking had suddenly fallen off a cliff, and panic, or despair, replaced the intricate probing of alternative approaches.

There are only so many pathways you can sketch out in your branching tactical diagram, and only so many dead ends you can come up against in the labyrinth of your thought, before everything blurs and foreshortens, your arguments go around in hopeless spirals, and the sense of self you'd lost in pursuing the problem returns with a stature stunted to suit this dreary perspective. You forget why you'd thought the question was worth answering, or why you had arrogantly assumed you could answer it. Knowing becomes nothing more than being told; it shrinks the world to a mere rattle of facts rather than layer on layer of structure, and the truth that shall make you free to an impersonal code of laws.

Dürer gave us the portrait of frustration in his engraving of Melancolia, and the English poet James Thomson put that portrait into words, concluding that the gigantic figure's eyes stared, only

> To sense that every struggle brings defeat
> Because Fate holds no prize to crown success;
> That all the oracles are dumb or cheat
> Because they have no secret to express;
> That none can pierce the vast black veil uncertain
> Because there is no light beyond the curtain;
> That all is vanity and nothingness.

Adolescent? Yes, but many of our students are adolescent, and there is a melancholy adolescent in each of us, moaning to come out.

How can you raise the threshold of frustration so as to stick with a problem for more than twenty minutes—and perhaps for the years it may take to follow its twistings and turnings and to corner it at last? Just as blood vessels proliferate around exercised muscles, so will channels of thought in neighborhoods of inquiry. The more we explore, the more ramified our explorings will become, and the more that external network and the inner network of our personality will merge. Like the characters in Ray Bradbury's science-fiction classic, *Fahrenheit 451*, who became the books they had memorized, experiment and experimenter, question and questioner, fuse.

A good example of this is an innocent-seeming question that leads into—and beyond—one of our Math Circle courses. Can you tile a rectangle with squares all of different sizes? What is especially interesting here is that intuition seems to nod neither toward affirmation nor denial, so that your thought very quickly tires from getting no nudges toward one possibility or the other. But you reach for a pencil and start drawing. Strategies grow, and are put on hold; they influence one another, evolve into subtler approaches, and generate broader and narrower questions—until each of the participants is wholly submerged in the enterprise and each other's point of view. This means that when the inevitable moments of frustration arise, there is a broad and intricate net of ideas, emotions, and language for them to fall into and rebound from.

Perhaps Heraclitus was right, that nature loves to hide itself, and what better disguise than skewed layers of translucent structure? But mind equally loves to uncover and is, after all, part of the nature it probes.

## The Jump-Cut Mind

The father of a seventh-grader in The Math Circle called one night to say he was worried. His son had become too engrossed in the problem we were working on. It wasn't like him, and it probably wasn't healthy for a boy that age to be so wrapped up in a math problem.

You could make an argument for the value of disordered, deficient attention (quickened responses to stimuli, a quarterback's agility in spotting momentary openings, unwillingness to follow paths walked on before), but the odds of winning the argument when applied to math are vanishingly small: math always seems to frisk about the edges of our understanding, and needs whatever lassoes and bridles we can find for holding on to it. The standards MTV has set for attention aren't ones we can measure math, or any of the arts, against.

We want to understand what Newton meant by saying: "If others would think as hard as I did, they would get similar results." One of the outstanding computer scientists of the last century, Richard Hamming, put it this way:

> I worked for ten years with John Tukey at Bell Labs. He had tremendous drive. One day about three or four years after I joined, I discovered that John Tukey was slightly younger than I was. John was a genius and I clearly was not. But Hendrik Bode disagreed: "You would be surprised, Hamming, how much you would know if you worked as hard as he did that many years."
>
> The more you know, the more you learn; the more you learn, the more you can do; the more you can do, the more the opportunity— it is very much like compound interest.

Is it pure athleticism, then? Will sheer plod make plow-down sillion shine? If it did, hitting a problem with the same blunt instrument for eight hours would crack it. Hamming continues:

> The idea is that solid work, steadily applied, gets you surprisingly far. The steady application of effort with a little bit more work, intelligently applied, is what does it. That's the trouble; drive, misapplied, doesn't get you anywhere. Just hard work is not enough— it must be applied sensibly.

Well and good, but what does it mean to apply your hard work sensibly? This involves capacities under the headings of Taking Apart and Putting

Together. Here the question is how to lengthen your span of attention without snapping or weakening it, so that it will support the strains of whatever those analytic and synthetic capacities turn out to be. Attention holds them in position as tautly as a trampoline's frame.

There are certainly exercises for lengthening your attention's span. Following someone else's reasoning step by step through a proof can be very helpful—especially if you have to make clear to yourself, along the way, what the tactical measures are and how the strategy has been planned. In a long proof the earlier parts may blur in your mind by the time you come to the later ones, and thinking or reconstructing builds them into your thought. The final test is always to run over the outlines of the proof in your mind. Could you explain it to your eight-year-old niece or your eighty-year-old grandmother? Will it hold their attention as it now fills yours? The drawback is that passively following another's reasoning may ill prepare you for hacking your way through the uncharted jungle of yours.

Two different sorts of factors, it seems to us, go into the lengthening of attention's span: a Rousseauvian eye and a Jeffersonian hand. That eye is full of the fire of commitment: you want to plunge into the problem and lose yourself in it. Heedless of others, reckless of your future—nothing else matters, so fully does it obsess you. This isn't an attitude you can fake, at least not to yourself. It is like falling in love. Time disappears in your engagement with this morsel of mathematics. (Time is in fact absent from mathematics as a whole, where we off-handedly invoke an infinity of ever-present instances with our "all", and by asserting "there exists" mean that something forever is.)

Simply swinging your attention around to the problem, immersing yourself in it without hesitation, will make causes appear within causes, and structures behind structures, as if you were following corridors in the pattern, invisible to the casual eye. You may joke with friends, take in a film, go on walking tours in the Cevennes, but the problem and its context is always with you, like the background hum of the universe. What in fact leads to your crossing the line that divides observer from actor? Something as remote from the problem itself as the personality of the person presenting it, or the excitement of those around you, may face you in the right direction. A question you ask, a suggestion you tentatively make, subtly alters your stance; then the endorphins kick in, and what was a pleasant annex of the day becomes its atrium. As in jogging, you have to go past the reading and talking and dreaming about it, the buying of shoes and clothes, and actually *jog*. And the second day may be harder than the first, because muscles ache not only in anticipation, but in reality. As in mastering a computer language, intimidation gives way all at once to an enormous feeling of intimate power. As in reading Shakespeare, the barriers of a vast, strange vocabulary and involved sentences are crossed again and again by lines that haunt you. Confidence and competence grow, feeding on one another. But whether intrigued or enticed, in the end it is your *will* that makes you a participant in the play rather than a spectator. This is never as simple as it sounds, since will has its own array of causes—and half the time, mathematics asks you to watch the behavior of its structures *without* imposing your will on them.

Altogether different, but as important, is the Jeffersonian hand. How manage your life, let alone this consuming struggle, without some sort of least governing? You know the answer from driving: how can you make the continual, minute adjustments to a car's steering wheel while clutching and shifting and braking and checking the side and rearview mirrors, not to mention the odometer, speedometer, fuel and temperature gauges and retuning the radio, all the while effortlessly maneuvering your way through the grammar of a sentence spoken to your companion, in which you decide in midstream to hold back this information, phrase that juicy tidbit discreetly, and finish with an ironical flourish, saying one thing but meaning another? At the same time, of course, you're squirming into a more comfortable position and readjusting the seat belt, while trying to remember if you'd put ketchup on the shopping list, and stray memories and half-acknowledged thoughts flicker and flow in the stream of your consciousness.

You bring off this paragon of complex and lengthy attention by having long since delegated its parts to routines that are now subliminal. While your spinal cord's lieutenants sort these matters out, the captain in your brain orders the thoughts that count—articulately enough for logic to clear the way, but with sufficient flexibility to respond to suggestive analogies. If it is a wilderness you are looking at, it is from the well-proportioned windows of an inner Monticello.

Best of all, this light governance of our thinking's network lets us become true travelers in it: lost at times, but never at a loss, because we have an underlying sense of our resources, if not of the geography. Fantastic visions may lead us around, but not astray; gratification, so long delayed, is replaced by the delight of small details. Your attention spans the problem and what you bring to it, as its personality becomes indistinguishable from yours.

## Putting on Hold

One of the charms of reading or listening to German is that a sentence may march positively along only to be brought up short by a *"nicht"* at the end. You need to reverse the signs of all your suppositions and pull up your expectations by the root. It's like one of those puzzle pictures that exchange figure and ground. It does teach you, however, to put in brackets larger and larger chunks of your thought, so you'll be prepared to see them from different angles or in altered contexts.

This willing suspension, not of belief but of commitment, is vital for mathematical expeditions. Think of the times your curiosity has been caught by something to do with numbers (do they really go on forever, although the universe may not? How does my calculator do its magic? What's the answer to this Sudoku puzzle?) You worry at it for a while, then let it go; it returns from time to time with diminished intensity, then fades away, to join the background mystery of things. To take the problem seriously would probably mean leaving the familiar world behind for months. You would need to practice thinking conditionally, which means intensely following premises through long and tortuous steps to an ultimate conclusion, but then recalling that they were, after all, only premises—and may not hold. If A then B, and so C and D. . . .: it was the "if" that drove Mr. Ramsay to distraction in Virginia Woolf's *To the Lighthouse*, and many a Mr. Ramsay before and since. If the Riemann Hypothesis is true, then look what follows: a looking that takes all your effort. But if we haven't yet proven the Riemann Hypothesis, you'll need all the effort you can collect in order to follow along, when you have only a promissory note in your pocket.

Perhaps learning this sort of committed reticence is the most unnatural call mathematics makes on our psychology. You find it at its most vivid in proofs by contradiction: let's see what follows if we take it that A is so (knowing all along that our effort is to show that A is not so). Now a novel's worth of consequences may follow before we discover that the end undoes our beginning. This isn't very different, is it, from the Hindu tale of the youth who encounters a god unawares and is asked to fetch him a bowl of water from the nearby river—and on the river's bank sees a beautiful girl and falls headlong in love with her, and they run off together, and marry, and have children, and live in a house by the river . . . until one night a flood sweeps the house and the children and his wife away, and nothing is left save the god, laughing beside him: "Where is my bowl of water? All is Maya!"

A mathematician reading a proof may catch less than half of its drift and less than that of its intricate argument. He isn't wasting his time, however, but getting a feel for where it fits in, and what it points toward. This keeps the details from blurring his vision—as scientists in their experiments hold all of the variables fixed save the one they wish to follow, and lawyers stipulate so as not to speculate. Of course, they keep auxiliary arguments stored up, and mathematicians, too, have neat packages of techniques and conclusions at hand, to take out and brush off when needed.

It is from the inner Monticello, once more, that we take control of our ifs and thens, and keep our minds clear to focus on the unfolding chain of consequences by storing alternative possibilities neatly away—ready, along with analogies and trusted techniques—to use when the time is right. We look out on chaos from our well-made window, and this keeps us from fretting: this landscape too we shall one day tame, and see beyond it the next frontier.

## 2. Taking Apart

You've plunged in and are thrashing about. As French mathematicians say, *"Ici on nage dans un bordel"*: literally, we're swimming in a bordello. Too much data, too few connections, questions swamping answers, every conjecture starting a dozen more with no direction of inquiry looking more likely than another. Even the terminology is losing its sharpness: the words have come to mean anything but what we wanted them to, both more and less (and you suspect that some may now mean nothing at all).

Take, for example, the question we mentioned on page 32: can you tile a rectangle with incongruent squares? There we are, caught on a cusp between yes and no. In Math Circle classes that have struggled with this problem, the first, failed flurry of suggestions often led to the strategic retreat of making more precise what the question was—and this is surely the beginning of taking apart.

"Put in the largest square you can—you know, with a side the length of the rectangle's shorter side," an eight-year-old in an otherwise adult London audience suggested.

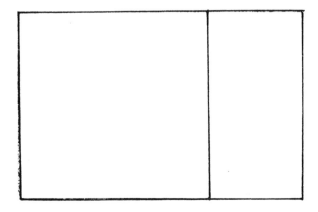

"Then do the same with the part that's left, and so on—"
"But what if the part that's left is longer than what you filled up?" someone else asked.

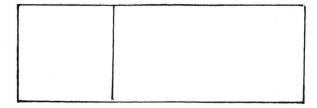

"Are you going to allow an infinite number of squares?" asked a third. We intentionally pose our problems vaguely, since a large part of doing math consists in making the initial question sharp enough to answer (while avoiding questions you don't know how to sharpen).

"What would you like?" we said.

"Well, it looks as if you might be able to do it with an infinite number of noncongruent squares, with side-lengths x, x/2, x/4, x/8, …"

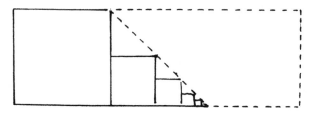

"Really?" said someone else. "Wouldn't that only fill up half of it—under the diagonal—and you'd have to do something different above, as well as filling in all those little gussets of triangles."

And another: "Even so, that would work only in 1 x 2 rectangles."

"You could always spiral around."

"But even then, would that always work? Did you say it should be for 'any' or 'some' or 'all' rectangles?"*

"And what do you mean by 'tiling'? Can the tiles overlap?"

"Or have grout between them?"

"Or can some be put in vertically?"

"Or some have a negative area?"

The question and the questioners exert mutual overt and covert pressures, and each deforms to fit the other: we refine our questions until we decide we've sufficiently well defined what they're about, and then—as we begin to answer them—refine and redefine more.

Here are some capacities that help in taking a problem apart to see what makes it tick.

---

*When we posed this problem to an audience of 500 at the Perimeter Institute for Theoretical Physics, in Waterloo, Ontario, one of those in the audience (the distinguished knot theorist, Louis Kauffman) came up in the course of the evening with a proof that such an infinite tiling works *only* if the proportions of the original rectangle are 1 × φ, where φ is the Golden Mean. See his paper, "Fibonacci Rectangles" (eprint: arXiv:math/0405048).

## Method vs. Approach

You need only look at the chaos of a child's room and hear his parents' nagging about it to see that being methodical is learned. The pleasures of a place for everything and everything in its place compete for dominance with less systematic delights. Necessity often mothers a reluctant sense of method in adulthood, when you find that just muddling through often ends without the "through". Sorting out the appropriate tools for attacking a problem goes some of the way toward solving it. So cooks let their creative tantrums loose on a carefully assembled *mise en place.*

But are there really "methods", like the ones promised by Get Rich Quick and Get Thin Fast plans? "Method" sounds to us like a form, independent of what it is stamped on—but in mathematics, at least, it helps to let the problem's topography guide you to the most direct or appealing ways into it, or to those that will make best use of your own strengths and viewpoints. Commonsensical advice, like "first decide what the question is," turns out to be more useful than such nostrums as "when in doubt, factor." Proverbial methods also have the problem of being applied indiscriminately, like the boy in the tale who, having dragged back a ham from the market and been told he should have carried it on his head, brought back the butter that way on the next hot afternoon. This is why in The Math Circle we tend to use words with less rigid associations than "method", such as "approach".

Say you want to prove the Pythagorean Theorem. You could look at this as an exercise in cutting up and moving polygonal shapes around, and some dissections and foldings or rotations of the pieces will come more readily to one person's mind than to another's. Or if your inclination is algebraic, you might translate some of these configurations into quadratics, or see the problem as centering on proportions. If you think like President Garfield, you fetch a trapezoid from afar to assist in your proof; if like da Vinci, you add two copies of the original triangle to the whole; if you are Einstein, you look at the area ratio of similar triangles. You might deduce the proof from the properties of chords intersecting in a circle, or from secant and tangent lines. You might apply the so-called "method of analysis and synthesis", reasoning backward from the conclusion to true premises, then following in reverse the route you made. You might even prove this theorem about a very small part of the plane by appealing to a limiting argument that involves the plane in its entirety. Why should we try to fit human ingenuity, bred of its bearer's personality, into the Procrustean bed of some sort of impersonal method?

Approaches, then: gloves that fit your hand as well as the problem's contours, the better to let you grasp it: a way of keeping tabs on what you've done and where it led, so that you can at least catalog the kinds of obstructions you came up against. Do they sketch out a pattern whose reflex shape might be a solution? A catalog too of analogies that flickered between this problem and others you've solved—or between parts of the problem and structures you understand. What often turns out to lead to a dazzling solution is generalizing the problem in one way or another, then looking at what sense it makes in that larger setting: keeping track, therefore, of the problem's evolution, and yours, and the evolution of their interplay, so that stepping back from time to time you may see revealing shapes.

Something as unimposing as cataloging may be the kind of thing Alfred North Whitehead had in mind when he recommended letting the infantry of your thought do most of the fighting, so that you can save the cavalry for the important charges. And as long as you don't take its categories too seriously, a list of what's worked in the past can help in the present: a tool chest that has deduction and induction in it, along with proof by contradiction, and by analysis/synthesis, and by turning your attention to your problem's contrapositive. Whenever someone devises a new approach to a problem, other people are quick to detach it from that problem and put it, oiled and wrapped, into this chest. More strength to your arm—so long as mind rather than habit or hope directs it, and mind and arm keep touch with the singular structure they probe.

## Atomizing

"Taking apart", you might argue, is just demotic for "analysis", and analysis, when you come right down to it, is nothing more than coming right down to it: breaking your problem up into its simplest parts and then dealing systematically with them, or reducing what you don't know to what you do—dissection rather than destruction. Those exploded drawings of engines and electronic equipment that used to be the favorites of mechanically minded teenagers, and now puzzle them as parents in directions for unclogging the washer, are exemplars of this sense of analysis, as are organizational flowcharts and stochastic diagrams. But is it always no more than a mindless matter of laying out the parts and then slipping tab A into slot B? Any problem worthy of the name always seems to have at least one stage that demands ingenuity. Were coming-to-know algorithmic, computers would be able to distinguish acquaintances from friends.

We used to think that if we knew one, we knew two, because one and one are two. We are finding that we must learn a great deal more about "and".—Arthur Stanley Eddington

The invention of the quadratic formula is a good example. Everybody had to memorize

$$x = \frac{-b \pm \sqrt{b^2 - 4ac}}{2a}$$

in algebra class, but you might have wondered how that magic answer was derived from the original question: what is x, if $ax^2 + bx + c = 0$?

You want to free *x* from its encumbrances in

$$ax^2 + bx + c = 0.$$

Easy enough to ship the constant across the river of equality,

$$ax^2 + bx = -c,$$

and to rid $x^2$ of its coefficient:

$$x^2 + (b/a)x = -c/a.$$

But now what? It took medieval time and Middle Eastern cleverness to come up with what we now blithely call "completing the square": adding $(b/2a)^2$ to both sides of this equation:

$$x^2 + (b/a)x + (b/2a)^2 = -c/a + (b/2a)^2.$$

That's the invention, which reducing called for but couldn't unthinkingly supply. And now it is all technique again—which is to say, reduction once more to routines you had long since mastered:

$$(x + b/2a)^2 = -c/a + (b/2a)^2,$$

*How does* $-c/a + (b/2a)^2$

$$x + \frac{b}{2a} = \pm\sqrt{\frac{-4ac + b^2}{4a^2}}$$

*get to* $\pm\sqrt{\dfrac{-4ac + b^2}{4a^2}}$ ?

and so to

$$x = \frac{-b \pm \sqrt{b^2 - 4ac}}{2a}$$

Reduction may, however, deceive you into thinking a proof will run on just the combustion you set going, without any further oversight or insight. Thus many a geometric proof begins by breaking a compound structure up into its constituent triangles—but then what? In his brilliant *Proofs and Refutations*, Imre Lakatos looked closely at Cauchy's proof that for a polyhedron with V vertices, E edges, and F faces,

$$V - E + F = 2.$$

Consider a cube, says Cauchy (in Lakatos's retelling), and remove the back:

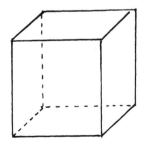

If our formula (the "Euler Characteristic") holds, then with one face gone, our new figure will have

$$V - E + F = 1.$$

And so we need only show this is true: the first appearance of reduction. Squash the open box (let the side pieces stretch, so no V, E, or F will be lost in so doing),

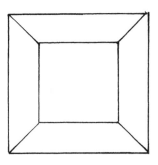

split each face into two triangles (reduction again, though of a different sort)—and notice that you have added an edge and a face to each of our original faces; but since E and F have different signs, there will be no change in $V - E + F$:    *E and F balance*

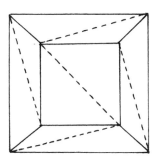

Then begin to remove these triangles (reduction's third appearance) in, say, this order:

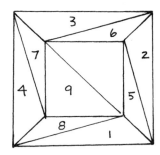

In removing each of triangles 1, 2, 3, and 4, a face and an edge disappear: again no change in V – E + F. In removing each of triangles 5, 6, 7, 8, and 9, the two edges that are taken away are made up for by the removed vertex and face. This leaves us (the last reduction) with one triangle, for which obviously

$$3 - 3 + 1 = 1$$
$$V - E + F = 1.$$

Add the original removed face back on, and

$$V - E + F = 2.$$

So, says Cauchy, the theorem is true for the cube, and the technique would generalize for a polyhedron with any number of sides.

Notice two quite different points. First, the tactic of reduction, applied again and again, was in the service of a strategy that arose from an idea that reduction alone wouldn't have come up with: Cauchy—or someone—had to hit on this ingenious analysis. Second, however, how do you *know* (ask the protagonists in Lakatos's dialogue) that you can do this remove-and-squash maneuver with anything other than a cube? How do you know that a different order of removing the faces of a cube will leave V – E + F unchanged? How do you know that—in some other polyhedron—you can always find a removal sequence which would leave V – E + F unchanged? Maybe, even if you were able to flatten out another polyhedron, you might create virtual edges, faces, or vertices, or obscure some of those that were there? And since we already know by simple counting that in a cube,

$$V - E + F = 2,$$

we didn't need this elaborate atomizing for the cube itself; it was meant only to exemplify a general procedure—but in fact it seems to exemplify no such thing. Although Cauchy's method may work each time you try it, he hasn't proved that it always works.

And what does Lakatos's analysis itself exemplify? Among other things, that simplifying may simplify the subtleties or the metamorphoses of a problem away; that atomizing, in the sense of breaking down, is a wonderful source of clarification, but must be used cautiously lest its light be blinding instead. Let your infantry indeed reduce the enemy's fortifications, but always under the direction of a distant commander, and always with your cavalry at the ready.

You can atomize the object of your thought into its constituents, or you can atomize your thoughts as well: reduce the problem to a simpler one (the first reduction in the previous example), or to one you already know (as with what's left of the quadratic, after completing the square). Mathematicians are notorious for this kind of reduction, as in what passes for a joke about the physicist and the mathematician who each want a cup of coffee. They go into the kitchen, the physicist fills a kettle, puts it on the stove, turns on the burner, and when the water has boiled, pours some into a cup with instant coffee. The mathematician then knows what to do: he empties out the kettle, thus reducing the problem to the previous case, and proceeds as before.

The point is that this kind of reductive thought, set about with so many advantages and perils, comes with the territory of being human: we all profit by and suffer from it. You need only think of how readily we reduce people and events into clichés to see how widespread the habit is. This is a case, then, not of developing an unnatural trait but of channeling one that is all too natural—the sort of laziness that always seeks out the path of least resistance. But don't dismiss laziness out of hand: efficiency experts look for a company's laziest employee, to see how he manages to do just enough to hold on to his job.

An outstanding contemporary mathematician, who believes as little as we do in any sort of talent, tells us that of all those he has met in the mathematical community, the one who most seems to have an inborn skill isn't in fact particularly imaginative, but has an uncanny willingness to break a question down into its tiniest details and then solve each one. This may come, he thinks, from this person having no native language *other* than mathematics.

You could therefore think of atomizing as a fundamental skill, itself atomic, which amounts to letting your thought, like water, follow the contours of what it flows over, seeking out the gradient, finding the fissures, and splitting the terrain apart along its natural fault lines. But doesn't atomizing itself really atomize into two skills: a diamond-cutter's sensitivity to those fault lines in a problem, so that a little tap will cleave it; and what we might call "accessible experience"—that repertoire of instances and approaches we spoke of, scanned now and again for something to fit the new configuration. So atomizing consists in breaking a problem, and how you approach problems, into their fundamental units; then laying one like a transparency over the other, and shifting them, until (with the click of revelation) form fits form.

Here's an example. Draw any closed shape you like on the plane—as simple as a pancake, or as rococo as this:

Is there a straight-line cut that will divide its area in half? Let's say that somehow or other you come up with this idea: a line lying wholly to the figure's left has all of the figure's area to its right. Likewise, a line parallel to the first, wholly to the figure's right, has all of its area on its left.

So were the left-hand line to be moved continuously toward the right-hand one, it must at some point divide the figure's area precisely in half.

This answers in the affirmative the question of whether there *is* such a line—although it doesn't begin to tell you where. A peculiar and somewhat disconcerting solution.

Now let's say that this "nonconstructive" answer not only sticks in your craw but in your memory, and years later you run across this problem: through a point P inside a convex shape on the plane ("convex" just means that the shape has no indentations—a line between any two points in it or on its border lies wholly inside it too), a line between two points, A and B, on the border is such that AP > PB.

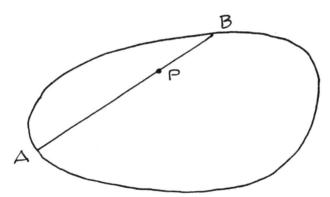

Is there any straight line through P, out to the figure's border at *some* A and B, such that AP = PB?   Yes.

The ghost of that previous problem comes to mind and you think: we began with AP > PB, but if we rotated the line through P until the positions of A and B were interchanged, we'd have AP < PB, so there must be some intermediate position where AP = PB.

This is a solution no more (and no less) satisfactory than that of its tutelary spirit: a style of proof, a kind of thinking, that once encountered, can generate answers to questions that have in common only the way they are posed.

**Attention Without Tension**

A quality we so much admire in athletes (the archer on a Greek pediment or the quarterback turning, faking, sliding out of the pocket, spotting his receiver in a chaos of oncoming aggression and hitting him with a perfectly threaded pass) is keeping a cool head when all about are losing theirs: poised potential, where readiness is all. This graceful harnessing of power lies at the base of taking apart, since it alone lets you analyze without becoming flustered by discordant data or discouraged by the shutting down of one path of inquiry after another.

Attention without tension sets you not so much in or above or around the problem as *at* it; it lets you understand by fitting your mind to the shape of what it contemplates: "unassuming, like a piece of wood not yet carved"; as a Zen master put it, "vacant, like a valley."

Say someone gives you this koan: in a large heap of coins, 877 are heads up, the rest are tails. Can you *blindfolded* divide the whole heap into two piles with the same number of heads in each? The easiest answer is "no." Yet if you turn your attention to it without letting your blood pressure rise, this and that begin to make themselves felt—as if indeed you were blindfolded and discovering the question's character by patting its contours. Two heaps with the same number of heads in each—hence an even number of heads; but 877 is odd. If the problem was correctly stated and there is a solution, it must somehow get around that difficulty.

When we posed this problem to a class of young Math Circle students, a thirteen-year-old suggested that you take all of the coins and, blindfolded, put them on edge—making in effect a roll of coins—then divide that roll into two parts however you like. Each will have the same number of heads showing: namely, none. This is a brilliant solution, arrived at by a combination of self-confidence, humor, and dwelling on the way the problem was stated: paying confident attention to each of its words and just laying out the range of their possible meanings. That "same number" could be . . . none.

Another student, thinking about it and "turning it this way and that in his powerful mind," as Homer describes Odysseus doing when baffled, decided that 877 was a red herring; there was nothing special about this number except that it was odd (and prime, which might turn out to play a role). Very well, simplify by restating the problem with an odd number that would be easier to deal with and visualize. Say the heap contained a single head and the rest tails. Now what? We must have two heads before we can even begin to do the separation. Notice the role that calm attention plays here: there will be another problem—the problem of separating the coins, blindfolded, into the required heaps—but that's for later.

Simply put it on hold for now without worrying about it, and first, some-how, get two heads.

Ah . . . no one said we couldn't flip a coin; so *that* would give us two heads—unless, by ill fortune, it was the single heads-up coin we ended up flipping. But then we'd have *no* heads, and so could now divide the pile up in any way, and each part would have the same number of heads: namely, none.

So it comes down to this: if it was some tails-up coin we'd flipped, we'd now have two heads: how guarantee they would end up in different piles? Confidence, bred of the "flipping" insight, propels relaxed attention along to the next insight: why not first move a single coin away from the heap, and *then* flip it! If it was a tails it is now heads, and so there is one head in each pile; but if it had been the single heads that was flipped, all are now tails, and each heap (the original pile, less one, and the "heap" of that single coin) has the same number of heads: again, none.

Will this approach work if we are told the heap contains two heads and the rest tails? Well, separate two coins from the heap, then flip each. If neither was heads, there are two heads in the pile, and now two heads in this two-coin heap, as desired. If both were heads, then after flipping, both heaps have no heads. If one was heads, one tails, there is one heads left in the pile, and flipping these two will reverse their faces, so one will be tails, one heads, as desired. Aha! 877 being odd played no role, since it works for two!

Does the reasoning change with three heads in the original pile? No, it just breaks down into four cases, each one of which works in the same way, *because the same principle is at work*: separate from the pile as many coins as there are heads in the pile; flip each of these separated coins— and this "change of parity" works like a charm.

Wonderful! And notice that attention without tension pervaded the thinking. For after the "flipping" insight it would have been very easy to become too excited to think calmly through the next stages—distracted by one's cleverness or, more selflessly, by the possibilities and questions opening up everywhere. A focus just narrow enough, then, not to be perturbed, yet sufficiently broad to let in the light that reveals the problem's texture: the restraints that weren't there, the latitude for in-venting that the terms allowed.

This receptive attention must also hold its focal object steady under the stresses applied to it by all sorts and degrees of ambiguity. In our problem, for example, it seemed that a number had to be both even and odd; and in any problem worthy of the name there will always be at least one of what the Japanese poets call *kiriji:* a word, a phrase, an idea that bears two senses liable to carry you off in different directions, like an open switch on a railway line—if it doesn't derail you altogether.

Ambiguity in mathematics? Many people summarize their objections to math by saying that it has none of the ambiguity that gives life and art their roundness and richness: the single possible answer glowers at us, insulting our deep love of freedom and taking from us the summer dream beneath the tamarind tree. In fact, mathematics seethes with ambiguity, which is among the reasons why, when you try to get to the bottom of any problem, you find it is bottomless. It is also why the great Georg Cantor proclaimed in the nineteenth century: "Mathematics is freedom!"

Our first, rough acquaintance with a problem inevitably leads to little ambiguities cropping up in how we understand it—or more insidiously, in our ignoring hidden constraints or seeing constraints where there were none (who said you couldn't flip a coin?). As we come to know it better, ambiguities signal our changing point of view. Picasso profiles emerge as the problem slowly rotates, and the longer we remain stymied, the profounder these ambiguities become: should our approach to it have been algebraic rather than geometric? Are we stuck in too conservative a two-valued logic, should we move from thinking of it linguistically to structurally? Should we be inventing rather than pretending to discover? Ambiguity segues into doubt, and the current shakiness in the foundations of mathematics undermines our thinking—can we gain insight without proof? Can we ultimately prove anything at all? Will it suffice to convince others—will we be able even to convince ourselves? The legacy of proofs that stood for so long, only to tumble eventually on closer scrutiny, hovers over our tentative work. And there's the puzzling self-referential quality of any piece of mathematics: where exactly are we and our little problem in the whole? Does the notion of a whole even make sense in this peculiar, infolded enterprise, or is it *essentially* ambiguous?

Certainly mathematics is pervaded by ambiguity, as the presence of equations everywhere in it shows. For what is an equation but the confronting of two different points of view? To say that the square on the hypotenuse is also the sum of the squares on a right triangle's two legs is the beginning, not the end, of a deep insight; to say that $-1$ is another name for $e^{i\pi}$, or that when thinking about polyhedra, the sum $V - E + F$ is also known as 2—is to glimpse startling unities behind the flicker of appearances. With appropriate ambiguity, the ambiguity that began by distracting us promises to be our guide.

Attention without tension amounts to walking steadily on through an inviting landscape, taking in its foggy valleys and cloudy peaks, pausing for views that seem to unify and views where everything falls or rises away. Like good explorers, we're willing to put up with a bit of uncertainty in our situation for the adventure of it all, and for the suggestiveness with which it frames our arriving.

*So does $e^{i\pi} = -1$. if the $i$ supplies the minus sign.*

Richard Hamming, from whom we've already heard, sees this question about ambiguity and attention from a window slightly canted to ours:

> There's another trait on the side which I want to talk about; that trait is ambiguity. It took me a while to discover its importance. Most people like to believe something is or is not true. Great scientists tolerate ambiguity very well. They believe the theory enough to go ahead; they doubt it enough to notice the errors and faults so they can step forward and create the new replacement theory. If you believe too much you'll never notice the flaws; if you doubt too much you won't get started. It requires a lovely balance.
>
> When you find apparent flaws you've got to be sensitive and keep track of those things, and keep an eye out for how they can be explained or how the theory can be changed to fit them. Those are often the great contributions. Great contributions are rarely done by adding another decimal place.

## Precision

Only a little out of true is false. This can be very discouraging in daily life; it can be paralyzing in mathematics. Apprentice carpenters are wisely told to measure twice and saw once—but measure twice, then take out a magnifying glass and measure again, and then use your micrometer? Truths grow in evasiveness with their growing profundity precisely because we haven't quite got their measure: because we see more or less what we need to see exactly. It is as if ever finer details of structure demanded matching refinements in our thinking. Safecrackers, they say, sandpaper their fingertips the better to feel the delicate fall of the tumblers. But mathematics goes past the niggling precision of third and fourth decimal places characteristic of the engineering it sprang from, to the glorious absurdity of infinite precision: each number taken to *all* its decimal places, each geometrical object honed to the ideal accuracy of thought.

A notorious example is that curse of first-year calculus students, the "$\varepsilon$–$\delta$ method". You want to say that as the inputs to your function f(x) approach a certain number, *a*, the outputs approach a certain value, *l*. Well, why not just say as much, and go on? Because the multiple ambiguities here aren't fruitful but pernicious, and the devil is hiding in these details to damn you. The beautiful understanding of change and motion that calculus seeks to achieve depends not on what happens at *a*, but near it—and this in turn requires us to make very precise sense of what "near" means: a sense that won't stumble over and into all sorts of unforeseen pitfalls (what if, for example, the function isn't even defined at *a*? We still want to be able to say that its limit exists, and is *l*. Moscow is still there, even though the three sisters never get to it). What 3 sisters?

A great many people worked hard, over a considerable span of time, to refine their intuition and ours into an impersonal way of speaking

*Is this a play by Chekhov?*

about and dealing with this idea of "limit". What they settled on is very clever, because it brings the free will and the personality of the explorer into this formal definition. If the function's outputs really do approach *l* as the inputs approach *a* (from either direction), then they will do so no matter how coarse or fine the degree of tolerance that you wish to measure by. If you ask only that the outputs be within an inch of *l*, then you will find inputs within a corresponding distance from *a*; if you ask for a tolerance of a hundredth of an inch, again the inputs will lie within a band around *a* that corresponds to this new restriction. In fact, for any positive distance ε you choose on either side of *l*, there will be a corresponding positive distance away from *a*—call it δ—that the inputs will fall within.

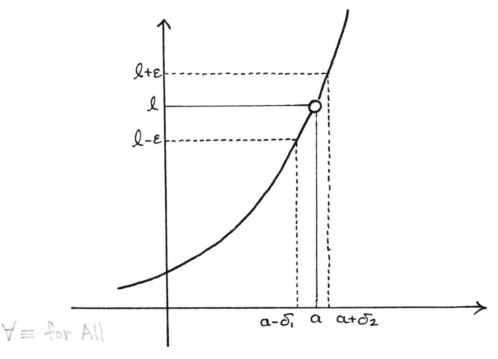

$\forall \equiv$ for All

Now watch how this just regard for precision, joined with an inclination to abbreviate, yields a piece of language that will scare not only horses: "For any positive distance ε you choose"—that is, for all ε; or if we use the symbol $\forall$ to mean "for all", this opening volley of our definition of "limit" becomes "$\forall\varepsilon$". Remember, though, that we wanted to say "for all *positive* distances away from *l*" (a subtle qualification: we're saying we don't care what happens exactly *at l*). So we should write instead "$\forall\varepsilon > 0$". For all ε greater than ∅.

What happens then? "There will be a corresponding positive distance, δ": so if we symbolize "there will be" or "there exists" by "$\exists$", this turns into "$\exists\delta > 0$". What we have so far, in our definition of "limit" is this:

$\exists \equiv$ Ther exists     There Exists a δ greater than ∅.

Definition: "The function f(x) approaches the *limit l* as x approaches *a*" means: "$\forall \varepsilon > 0, \exists \delta > 0 \ldots$"

And now what? We want to say that as long as the input, *x*, is within $\delta$ of *a*, then f(x) is within $\varepsilon$ of *l*. Well, "within"— i.e., no more than $\delta$ away from *a*, on either side of it—is nicely caught by the absolute value notation, which expresses distance: so "the input, x, is within $\delta$ of *a*" turns into "The distance between x and *a* is less than $\delta$", or:

$$|x - a| < \delta.$$

Likewise, "f(x) is within $\varepsilon$ of *l*" becomes

$$|f(x) - l| < \varepsilon.$$

Now we can put all these very fine bits of crocheting together to arrive at our perfected definition of limit:

Definition: "The function f(x) approaches the *limit l* as x approaches *a*" means: "$\forall \varepsilon > 0, \exists \delta > 0$ such that if $|x - a| < \delta$, then $|f(x) - l| < \varepsilon$."

If, in the full frenzy of abbreviating, you decide to ditch the mere English words "If ... then ... " and replace them by the symbol "$\Rightarrow$", and scrap the phrase "such that" and put "$\ni$" in its stead, then you can reduce the sense of our definition to a runic:

*What is this $\ni$? Eh?*

Definition: "$\forall \varepsilon > 0, \exists \delta > 0 \ni |x - a| < \delta \Rightarrow |f(x) - l| < \varepsilon$.

We have agonized with you over this condensation of so much watchmaker's tinkering in order to savor the spirit that hovers over it—the caution and its rewards. A precise grasp of what the notion "limit" entails is mirrored in the notation, which reflects its standards of accuracy back into the way we henceforth think about this notion. Once having stepped onto the beam, you have to go on keeping your balance; a little mistake will bring everything down (you can't, for example, reverse the order of "$\forall \varepsilon > 0$" and "$\exists \delta > 0$": if you thought that the definition of limit began "$\exists \delta > 0$ such that $\forall \varepsilon > 0$", you would be talking about something significantly different). You might get away in conversation with saying "They all didn't do it" when what you meant was that they didn't all do it, but the bones of mathematics are as delicate and as linked as those of a bird: twist one and the whole lithe being falls to earth.

This doesn't mean that in order to take a mystery apart you need to turn into a Dickensian bookkeeper, grown gray in the service of parched accuracy. Any craftsman will tell you that this kind of care, this attention to the meshing of barely visible gears, isn't only exhilarating in itself (and a sort of haven in the midst of the world's clamor), but the frame within which invention prospers. Think of the years spent coordinating eye and arm that let a great pitcher shave the slenderest edges off the strike zone.

Precision allows conception to become reality. In the nineteenth century, Charles Babbage designed the first machine that could solve simultaneous equations in several variables: his Analytical Engine. The best craftsmen of his day were unable to build it from his drawings, and people then and after concluded that it was conceptually flawed. A century later, working to far finer tolerances, technicians at the Science Museum in London built a model that whirred and hummed out its solutions: as satisfying in its brass workings as our image of what H. G. Welles's Time Machine should have been like.

**Rotating the Diamond**

If you look from a train window at a passing orchard, the trees are just a jumble, until the one instant when they all line up in neat rows—before the train moves on and all's chaos again. This is the moment you look for in mathematics, when turning a structure around suddenly reveals the hidden symmetries in it. Taking apart can amount to no more than changing the relation between you and your problem.

Richard Hamming knew that Bell Labs couldn't hire enough assistants for him to program computing machines in absolute binary, but rather than go to a less interesting place that could, he said to himself, "Hamming, you think the machines can do practically everything. Why can't you make them write programs?" And so, he says, he turned a defect into an asset, and observes that scientists, unable to solve a problem, begin to study why it might be impossible to solve—and get an important result.

Here is a representative case, from a Math Circle course called "Interesting Points in Triangles" (its format is described in chapter 9). The students first struggle to prove that the perpendicular bisectors of the sides of a triangle are concurrent—that all three meet at one point.

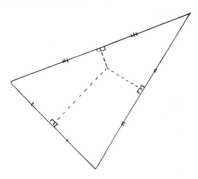

Then they are led to look at the concurrency of angle bisectors,

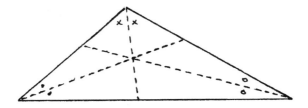

and then of medians, which run from a vertex to the middle of the opposite side.

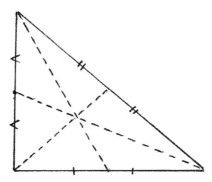

Inventing the different sorts of proof each of these require will have left the proof for perpendicular bisectors—and even the fact of their concurrency—buried in a communal subconscious. When they are led to ask

A mathematical point is the most indivisible and unique thing which art can present.—John Donne

about the concurrency of the altitudes—another set of perpendiculars, after all, though this time dropped from a vertex to the opposite side—

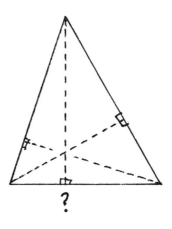

?

might they wonder if one triangle's altitudes could be the perpendicular bisectors of another? Turning the problem around—or inside out—they take the triangle ABC, whose altitudes they wish to show concurrent, and draw through each vertex a line parallel to the opposite base and twice its length, with that vertex as midpoint. Then a new triangle, DEF, takes shape, whose perpendicular bisectors are indeed ABC's altitudes. Since the former concur, so must the latter.

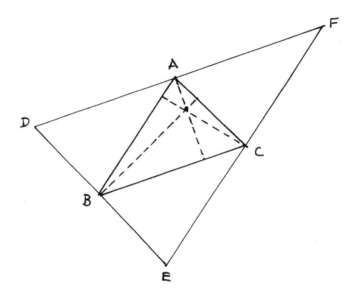

As with everything it touches, mathematics takes even its approaches, such as this, and turns them into its content too. "Seeing from another standpoint" becomes "duality": that peculiar yet characteristic undertaking where opposites exchange roles. You see this most vividly in projective geometry, where what were points become lines and vice versa; *duality!* and in the equations that describe them, variables and constants reverse. This outrageous act of the imagination sheds light over the whole projective plane—and over the mind as well that contemplates it.

Rotating a diamond-hard problem around is an antecedent to ingenuity. It allows you to shine the full light of your intelligence, and to bring to bear all your stored-up experience and powers of analogy, on facets previously hidden. We may puzzle over the physicist's aphorism that in quantum mechanics, the experiment and the experimenter affect one another; but we can't doubt that in math it is almost impossible to distinguish the subject from the object of thought: for the rotated problem likewise rotates our way of thinking, as if the two were geared to each other. And isn't the structure that mathematics contemplates the structure of thought?

## Breaking Apart

Why should "Breaking Apart" be a subheading of "Taking Apart"? Because of the violence involved. It's one thing to lay out your shoes each evening with the obsessive neatness of the little girl who was to become Queen Elizabeth II, and quite another to make candies fly from a piñata by hitting it as hard as you can with a bat. Taxonomy lets us understand a complex landscape by imposing order on it, if none springs naturally to view: no Darwin without a prior Linnaeus. But a stifling order (the regularity of unquestioned conventions) leaves you with nothing but dead certainties. That's the time for boisterous iconoclasm.

Everyone knows Euclid's parallel postulate (in the nineteenth-century form of Playfair's Axiom): through any point not on a given line, there is one and only one line parallel to the given line—

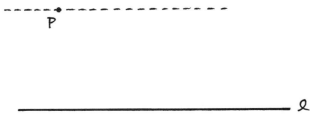

and thinks it is called a "postulate" for some arcane reason, since it is obviously true that there can be only one line parallel to another through a point not on that line. And yet those bold enough to discard this postulate and replace it with an opposite (many lines through the point parallel to the given line; or none at all) founded beautiful geometries—as consistent, as true, as Euclid's.

Everyone knows that it makes no difference in what order you multiply two numbers: ab = ba. Yet Hamilton came up with his astonishing quaternions when he was willing to break this law. Cayley abandoned the even more fundamental associative law (that a(bc) = (ab)c) in devising the numbers that bear his name.

Anyone but a crank knows that a proposition is either true or false. "A shade, they say, divides the false and true"—but what is this shade that Omar Khayyam says divides them? You'd have to be really iconoclastic to begin thickening a shade. And yet there are those who experiment with three- and, in fact, multiple-valued logic; the constructivists rightly question the status of existence proofs founded on contradictions—and what about the ever increasing number of propositions proven to be undecidable?

We're speaking not of the foolish iconoclasm that questions everything for the sake of sounding clever ("What if it were the other way?")

but of the often anguished refusal to accept what just doesn't make sense to you. The Russian mathematician Nikolai Nikolaievich Luzin, unable as a student to swallow the notion of "limit" as it was served up by his university professors, wrote of "intense intellectual suffering and pain," adding: "and the more intense the suffering, the better, for suffering is the source of creativity." Perhaps—but breaking open the box to discover how the locking mechanism works will leave you for a long time with a broken box.

Luzin tried to explain his ideas to his professor, who ended up storming at him: "Always the same! I am talking to you for half an hour about *limits* and not about your actually infinitely small which doesn't exist in reality. I prove this in my course. Attend it—although for the time being I don't advise you to do so—and you will be convinced of this. . . . Is there anything else you want to tell me?" Rather than saying that this was no way to treat a student, Luzin just walked away. That he persisted on his own in rethinking infinitesimals through the next two years, and indeed, throughout his whole adult life of mathematical research, epitomizes the stubbornness we spoke of earlier. It should come as no surprise that our categories aren't disjoint, and that holding on and tugging apart are only different incarnations of Beau Geste, firing off cannons from here and there along the parapets of Fort Zinderneuf.

A healthy iconoclasm not only frees you from too timid a relation to authority ("The systems are rot," said T. E. Lawrence) but supports the buoyant experimentalism whose motto is Nietzsche's: "Let's try it!" The limit toward which these complementary processes converge is the Themistoclean spirit: with the Persians about to invade Athens by sea, the walls to the port of Piraeus had to be strengthened—and where was the stone to come from? Throw in the statues from the Acropolis, said Themistocles. Destroy our greatest works of art? It's that, or lose to the Persians—and if we win, we will carve yet more glorious ones.

If you object that "iconoclasm" overstates the case, that it is as important to attend receptively to a problem as it is to peel away the hardened layers of unexamined assumption that obscure it, we agree. The qualities we have been looking at in isolation not only overlap but need to be titrated with one another: and (in the self-referential style of mathematics) this titrating is itself a quality, which we shall soon look at. Here let's compromise by saying that doing mathematics subtly changes your attitude toward *permission*. In his wonderful book, *Imagining Numbers (In Particular, $\sqrt{-15}$ )*, Barry Mazur tells the story of Gabriel García Márquez first reading Kafka's *Metamorphosis* as a teenager and falling off his couch, "astonished by the revelation that you were *allowed* to write like that!" You're allowed to invent a number whose square is −1? You're allowed to

*Luzin and infinitesimals. I'm with Luzin.*

pretend that an angled path is straight? *Versuchen wir's!* Try it and see what happens.

**Pursuing the Possible**

Of all our inborn analytical tools, the one we most enjoy is deduction, that pursuit of the possible as it scurries simultaneously down divergent branching pathways. We like it so much that we read detective stories in our spare time and elevate to the upper echelons of society those whose work centers on the deductive—doctors, lawyers, and engineers.

Astonishingly few statements in mathematics assert that something *is* ("There is a number, 0"; "There is a number, 1, and 1 ≠ 0"). The vast majority much more modestly claim that *if* such-and-such is the case, *then* something else must follow. This vast hypothetical structure rests, like St. Mark's in Venice, on the slenderest pilings, shallowly anchored in Being.

A dog with a long stick in its mouth comes to a narrow opening in a fence, stands back, stares, tilts its head sideways and trots through. A squirrel in winter, eager for the seeds in the capped bird feeder, hangs upside-down and unhooks the wire holding the cap on. We're not alone: animal life and deduction seem made for one another. Then why does it take so much trained effort to follow, and so much more to produce, the deductions in mathematics? The slenderness of those spiles is largely to blame: we're often so far from familiar ground that we feel none of the solidity that intuition, or just belief, supplies: deduction, for all its authority, seems not quite able to authenticate itself. Why else do we speak of being "forced" to a conclusion? Why is our science fiction populated by coldly rational aliens and inimical thinking machines?

Since mathematics, music, and mountains are, as we remarked before, often enjoyed by the same people, let's look at the latter two for some help in better understanding the effort needed here. A kind of backward inevitability suffuses great music. When you begin to listen to a new piece, the only sense you have of where it is going is the confidence you have on hearing a spoken sentence: confidence that the enveloping grammar will carry it safely, through however many adventures, to a conclusion whose shape will grow progressively clearer to you. Once heard, however, the continuous fit and flow of part to part makes you feel it couldn't have been constructed in any other way. So it is with following an intricate proof: why is this here, and that there? Which implication will we be following at this juncture, and where will it take us? Only long afterward do we see the design of the whole and recognize the intricate inevitability of the reasoning. Holding on to a strand of melody, a cadential sequence, is at a premium in these inner toils.

Holding on: those skills we spoke of before come sharply into play when you're struggling up the mountain of someone else's exposition. The preliminary demonstrations, the lemmas, are pitons driven into the rock face—and at a height where it is hard to catch a reflective breath. Although vast views surround you, you must keep your attention fixed on the minutiae of saving crevices.

Since deduction is meant to display, cogently, from premises to conclusion, the design of what we seek to understand, and since our minds are tuned to just this sort of display, why should there ever be any difficulty? Because our imagination frisks ahead of us at every turning and suggests different directions we might have headed off in; because we have to keep so many "ifs" in mind as we clamber from one "then" to another; because the techniques of deduction may change along the way (an inductive argument, a little contrapositive, an episode of proof by contradiction, in the midst of the on-linking network); because you must keep up the pretense that the exposition is impersonal, when in fact all the little leanings of the author's personality led to seeing an opening here, and chose the direction of ascent there; because chasms will open up that you must get across—technical passages in an unfamiliar technique—and you take the leap on faith now, but must promise yourself to build a bridge later.

These are just the problems of climbing established routes, laid out by others! What happens when it comes to taking a problem apart yourself, with deduction's aid? You have to be careful, at each of your argument's links, that the reasoning is sound. For while deduction as a strategy is straightforward, the growing sophistication of its tactics means that unobtrusive errors can creep in. Even with Aristotelian syllogisms you had to beware of an undistributed middle (people are either rational or irrational; real numbers are either rational or irrational; so real numbers are people). All those scrambling quantifiers have to stay roped together and, with negation, have to arrange themselves as their leader directs, lest they tumble out of control. Worse—it is all too easy for unwarranted assumptions to slip covertly in, under the camouflage of implication. The remit of a word or a concept may grow subliminally, making the transition from premise to conclusion deceptively smooth.

The real problem, however, is that so much follows from so little. We only pretend that deductive thinking is linear; such a wide splay of conclusions opens out of any premise that we often don't know which path to take or even where we might be taking it to, so that reasoning begins to circle about and can end at a standstill ("Mind," by Richard Wilbur):

> Mind in its purest play is like some bat
> That beats about in caverns all alone,
> Contriving by a kind of senseless wit
> Not to conclude against a wall of stone.

This may have been true of Russell. It is certainly not true for me.

You can even come to believe (as Bertrand Russell did, late in life) that all mathematical thinking is tautological, giving an illusion of insight but in fact only saying the same thing in different ways. This conclusion follows from too great a belief in deduction itself. But deduction is only a means toward taking apart: to *see*, by putting together, takes those kinds of imaginative play for which holding on and taking apart have set the stage.

## 3. Putting Together

The famous Banach-Tarski Paradox of set theory shows that you can cut a pea into a finite number of pieces and put it together again to be larger than the sun. How the mind fits its scattered rubble of questions, its ordered shards of insight, into the whole of a meaningful answer is almost as surprising. What began so recently as serious doubt becomes playful guessing, rearranging the parts that didn't fit together so that they will: even if this means impudently deforming them—or the rules. Must you really play the hand you've been dealt?

**Play**

Here is an example so beautiful and so telling that it's worth the effort it may occasionally cost to follow (at least this will make you sympathize with the audience at a math lecture).

Working on a problem rather like that of finding Pythagorean triples, the Swiss mathematician Leonhard Euler needed to know which natural numbers $x$ and $y$ satisfied $y^3 = x^2 + 2$. None, perhaps—or a single pair, or a few, or many—perhaps infinitely many. How could anyone possibly tell?

Reduce what you can't manage to what you can. In the classic technique of first-year algebra, factor: reduce the quadratic on the right hand side to its linear factors. The only difficulty is that the right-hand side *has* no factors, since factoring, by definition, requires integers: your only possibilities would be $(x + 2)$, $(x - 2)$, $(x + 1)$ or $(x - 1)$—none of which work. $(x + 2)$ times $(x + 1)$, for example, gives you $x^2 + 3x + 2$, and every other pair you choose will also produce an unwanted middle term. Nice customs curtsey to great kings, says Shakespeare's Henry V, and common definitions must likewise curtsey to stubborn mathematicians. Euler wanted this expression factored, so he factored it into

$$(x - \sqrt{-2})(x + \sqrt{-2}),$$

for all that $\sqrt{-2}$ isn't an integer. "I didn't know you were allowed to do that!" we might exclaim, leaning over the eighteenth century from our perch in the twenty-first.

Having gone so far, Euler saw nothing wrong in going farther. He decided that these two factors themselves had no common factor (which seemed to him a very reasonable decision: for what could possibly divide both?). Yet their product was, as we already knew, a cube:

$$y^3 = (x - \sqrt{-2})(x + \sqrt{-2}).$$

Since the two terms have no common factors, we can't combine, say, a couple of 2s from one with a 2 from the other to make a $2^3$. Each factor must, therefore, be *itself* a perfect cube. Now a perfect cube in the realm of complex numbers is simply some complex number, cubed. This means that $(x + \sqrt{-2})$, for example, must be some $(a + b\sqrt{-2})^3$, where $a$ and $b$ are integers. Likewise, $(x - \sqrt{-2})$ would be some $(c + d\sqrt{-2})^3$.

Pause for a moment to catch your breath: to find all *real*—in fact, *integral*—solutions of $y^3 = x^2 + 2$, Euler has audaciously rewritten his equation with complex factors, each of which must be a perfect cube.

Let's take one of them—say $(a + b\sqrt{-2})^3$—and actually expand it fully:

$$(a + b\sqrt{-2})^3 = a^3 + 3a^2b\sqrt{-2} + 3ab^2(-2) + b^3(-2)\sqrt{-2},$$

that is:

$$(a + b\sqrt{-2})^3 = (a^3 - 6ab^2) + (3a^2b - 2b^3)\sqrt{-2}.$$

And since he'd defined $x + \sqrt{-2}$ as $(a + b\sqrt{-2})^3$,

$$(x + \sqrt{-2}) = (a^3 - 6ab^2) + (3a^2b - 2b^3)\sqrt{-2}.$$

When you have an equation with complex numbers, such as this one, with real and imaginary terms that don't combine, the real part on one side must equal the real part on the other, and the imaginary parts will likewise be equal. Here this means that

$$x = a^3 - 6ab^2,$$

and

$$\sqrt{-2} = (3a^2b - 2b^3)\sqrt{-2}.$$

Dividing both sides by $\sqrt{-2}$ gives us

$$1 = 3a^2b - 2b^3,$$

or (factoring out the b),

$$1 = b(3a^2 - 2b^2).$$

Step back and look at what this says (and notice that this stepping back is as creative a move as all those that preceded it): Euler had factored 1 into the product

$$1 = b(3a^2 - 2b^2).$$

Now we begin to return to the natural numbers from our long excursion into the complex, and go back to our familiar definition of factors as integers, which multiply together to produce the desired product. The only integral factors of 1 are 1 and −1. If we let b = 1, then $3a^2 - 2 = 1$. $3a^2 = 3$, $a^2 = 1$, so a is either 1 or −1.* We're almost there. Remember that

$$x = a^3 - 6ab^2,$$

so if a and b are both 1, then x = 1 − 6 = −5.

But if a = −1 and b = 1, then x = −1 + 6 = 5.

Since −5 isn't a natural number, the only natural number value for x is 5. Replacing x by 5 in $y^3 = x^2 + 2$, we find that

$$y^3 = 25 + 2 = 27,$$

that is: y = 3. Trial and error would have shown that 5 and 3 do work; but by the most daring and playful series of maneuvers—and some assumptions only justified by much later work—Euler concluded that the *only* solution to $y^3 = x^2 + 2$, in all the infinity of natural numbers, is x = 5 and y = 3!

What Euler cooked up here, in his creative enthusiasm, probably took more holding on than we had to endure in following him, and a long taking apart of the notions "factor" and "complex number", until he understood each with the easy familiarity that let him risk and invent as he chose. The distance that time has interposed, and the even greater distance between lines of life, make what he did seem to have come up as things do in play—just playfully.

---

*b can't be −1, for if it were, $3a^2 - 2b^2$ would also have to be −1 (since −1 times −1 equals 1). But this would force a to be $1/\sqrt{3}$ or $-1/\sqrt{3}$. Were a to be either of these, however, with b = −1, then x = $a^3 - 6ab^2$ would not be a natural number, as we require it to be. So b must be 1.

### Stubbornness Revisited

*Their style of writing is a lot like Philip Morrison's*

It was stubbornness that let you hold on long enough to reach this creative height, where now flexibility rather than rigidity matters. But is some stubbornness yet needed—stubbornness suited to these changed circumstances?

"I will not let thee go except thou bless me," said Jacob to the angel as they wrestled all night. Wrestling with a problem for what may be much more than a night will take coming to grips with it in one cunning hold after another, until the problem blesses you at last with an answer. Mere holding on has evolved into shaping yourself to the problem's nature, so that you know it well enough to anticipate how it will next try to evade you—and you are there, waiting. Knowing what doesn't work edges you toward what might.

An earlier generation of mathematicians, comfortable in a language that took for granted a sharp distinction between the conscious and unconscious mind, thought of this long struggle as consciously constructing various combinations of ideas—which were then somehow reformed in the unconscious, to emerge into the light of day as a solution, aided by chance in those depths and by choice in our waking. Perhaps; or perhaps these two phases of mind are so fractally interpenetrated as not to be two phases at all, but a virtual continuum. The "combinations of ideas" flicker in and out of focus, like wrestling grips smoothly and suddenly transforming from one into another; and judgment all along appraises what hasn't yet—what still might—succeed; until the many parts of mind, and of the problem, grow unified. Or we acknowledge it to be a body too slippery for our grappling.

This aspect of putting together, then, is a negative one: building up a picture of what the problem's nature is by sketching out what it isn't ("By indirections," Polonius tells Laertes, "find directions out."). Remember that problem about the heap of coins with 877 heads in it? The problem *wasn't* about 877; it wasn't about prime numbers or odd numbers. Remove all this extraneous matter, and the idea of *changing* the states of some coins may glimmer through. In order to get this far with a stubborn problem—and any problem you haven't solved is stubborn—you need to fall in love with working on it, and this is obsessive love that doesn't hesitate to stalk, to pry, to probe. Even more psychologically extreme, you need—as we said before—to throw the whole of your personality into it, at the risk of becoming enthralled.

What happens once the problem is solved, the insight gained, the theorem proved, the theory understood? Different people react differently, as they do in affairs of the heart. With an English taste for depressive irony, the number theorist J. E. Littlewood wrote: "When one has finished a substantial paper there is commonly a mood in which it seems that there is really nothing in it. Do not worry. Later on you will be thinking 'At least I could do something good *then*.'" They think of these things differently on the continent: an Italian mathematician built for himself a triumphal arch in his Rome apartment and walked slowly through it, garlanded, whenever he proved a theorem. But the French mathematician André Weil said that "one achieves knowledge and indifference at the same time." *That seems like him.*

## The Riddle of the Pygmy Shrew

Some small mammals, like the pygmy shrew, have to keep eating for dear life; they can hardly take off the time to scramble from one source of food to another. When you get wrapped up in a mathematics problem you often feel the same way—don't stop me to ask where I'm going, it's taking all my concentration just to go there! One conjecture has led to another, which needs a lemma to support it, and that entails a calculation, which in its turn requires a digression to justify it, while the insight you were after recedes into the distance.

A key part of putting together consists in stepping back at judicious moments to take in more than the next turning. Such pauses not only calm the panic brought on by deduction's superfluity of consequences, which hide the shortest path to your goal—but counteract the ever-present infantile hope that plowing on blindly will somehow lead you to the solution.

You saw an example of this stepping back when, with Euler, we looked to see what it *meant* that $1 = b(3a^2 - 2b^2)$: it meant that each of those factors must be 1 or −1, and that insight (or far sight) led swiftly to what he sought.

Since that stepping back was a small part of Euler's overwhelmingly playful daring, we could do with an example that features stepping back as a major force for insight in its own right. Let's prove that $\sqrt{2}$ is irrational by assuming it to be rational and hoping for a contradiction. We assume, that is, that $\sqrt{2} = p/q$, where p and q are whole numbers with no common factor (otherwise we could reduce the fraction before going on). The proof will now flow in deductive style, guided only by the desire to simplify as much as possible.

We assume that $\sqrt{2} = p/q$. The most complicated part of that equation is the ungainly square root, so we do away with it by squaring both sides:

$$2 = p^2/q^2.$$

What's most complicated now is the fraction, so we continue to simplify by multiplying both sides by $q^2$. This gives us

$$2q^2 = p^2.$$

There's not much to simplify now, so instead of hunting around for another deduction, this is the time to step back and ask what we've learned. Well, for what it's worth, we see that $p^2$ must be even: it is 2 times some integer. A small aperçu, but the whole way forward will turn on it.

We now follow our noses again: $p^2$ is even, so p must also be even (for an odd squared is odd—pause for a moment to prove that—or glance at the footnote.)* Let's say p = 2a, for some integer a. Of the many paths we might have taken, the desire to simplify, and a small observation made on stepping back, have led us to this. We continue to follow our noses by substituting 2a for p back in the equation $2q^2 = p^2$. This gives us:

$$2q^2 = (2a)^2.$$

Simplify again (stepping back for a methodological moment, you might consider how far the desire just to neaten up has taken us):

$$2q^2 = 4a^2.$$

And of course, keep simplifying:

$$q^2 = 2a^2.$$

We are at the same kind of junction as before, with nothing much of a simplifying sort to do, except to step back: $q^2$ is even, so q must be even too. But wait: p was even, and now q is, so they must have the factor 2 in common, contradicting our assumption that p/q was in lowest terms! Hence $\sqrt{2}$ can be rational only if a fraction can both be and not be in lowest terms.

It takes a while for the full weight of what we've proved to hit you: there are perfectly ordinary square roots that aren't rational! It may take even longer to be struck by the beauty and elegance of the proof, with its plunging ahead and recoiling to contemplate where we were: a little fuguetta on how the mind synthesizes.

We can step even farther back from stepping back to see what a central role it plays in mathematical thinking. Time and again the way toward understanding a proposition, and proving it, consists in first seeing it as particular, then turning to the general context it belongs to, and doing your proving there. The irrationality of square roots provides a good example.

---

*An odd number is one greater than an even number: 2n + 1. Square that and you get $4n^2 + 4n + 1$, which is an odd number.

To prove that $\sqrt{3}$ is irrational we could proceed much as with $\sqrt{2}$, but now argue not in terms of odd and even but of the three possible forms that an integer may take: it may be a multiple of 3, like 3a, or such a multiple plus one (3a + 1), or plus two (3a + 2). 3a + 3 is back to being a multiple of 3. The assumption that $\sqrt{3}$ is rational would then involve this threefold distinction in the same sort of maneuvers we saw for $\sqrt{2}$. And the irrationality of $\sqrt{5}$? Now we would reason similarly, but with the five possibilities: 5a, 5a + 1, 5a + 2, 5a + 3, and 5a + 4. So it would go for the square roots of all primes, though extra work and some different turns of thought would be needed in order to show that $\sqrt{6}$, for example, was irrational. No wonder that Plato's tutor, Theodorus the Pythagorean, was renowned for having shown (by a separate proof for each) that $\sqrt{3}$, $\sqrt{5}$, ... up to $\sqrt{17}$ were irrational!

Must we really go through such contortions again and again? What weariness would proving the irrationality of $\sqrt{1848}$ involve us in? Step back. Let's prove at a stroke that $\sqrt{n}$ is irrational if n isn't a perfect square. We won't rummage about in the many different forms that numbers may have, but think on a more general level.

We need to know (and this takes a bit of proving) that if a prime number divides the product of two numbers, then it divides at least one of those two: if 1848 can be divided by 2 (and it can, because it's even) and if 1848 = 7 × 264 (and it does), then either 7 can be divided by 2 or 264 can be (which is true, because it's even). This points you toward the fact that any natural number can be uniquely decomposed into a product of primes. 1848, for example, ends up as $2^3 \times 3 \times 7 \times 11$, whether you begin with 2×924 or 3×616 or even 11×168.

With this knowledge in hand, assume that $\sqrt{n}$ = p/q (again, with p and q having no common factor), and as before, square both sides:

$$n = p^2/q^2.$$

We now have an integer, n, equal to a fraction, p squared over q squared. p/q was in lowest terms, i.e., p and q had no common factors. What about p squared and q squared? They have exactly the same factors as p and q had, except double copies of each; so if p/q was in lowest terms, then p squared over q squared must also be. Wait a minute: if the integer n is equal to a fraction in lowest terms, that can only be a fraction whose denominator is 1; so n = p squared, and so is a perfect square. But that contradicts our hypothesis. We have thus proven that if its square root is rational, then n is a perfect square, so if n isn't a perfect square, its square root is irrational.

There is something rather awesome in this capacity of mathematics to sweep through whole continents of instance in a single glance, retreating from the particulars while preserving what matters about them. Awesome too is the way taking one step back always seems to invite another. Shall we prove that the cube roots of non-perfect cubes are irrational, and the fourth roots of non-perfect fourths too, and in fact, that the mth root of any number which isn't a perfect mth power must be irrational? It will be the work of a moment.

For assume that $\sqrt[m]{n} = p/q$ (again, with p and q having no common factor). We take the same first steps as before, raising both sides to the mth power:

$$n = p^m/q^m$$

We now repeat, word for word, our previous argument, but applied to the exponent m rather than 2: p to the m and q to the m have nothing other than m copies each of the factors of p and q. But since p and q have no common factor, this is impossible unless q = 1. But if q = 1, we have n = $p^m$, i.e., n is a perfect mth power, which contradicts our hypothesis. We have thus proven that if n isn't a perfect mth power, its mth root is irrational. We could step even farther back. . . .

Algebra is generous, she often gives more than is asked of her.—D'Alembert

Putting together is almost indistinguishable from stepping back, so that learning how and when to do it (without falling off the narrow ledge you've climbed to) is central to doing mathematics. A mathematician turns out to be more like a contemplative gibbon than a tree shrew.

## Analogy

"Analogy is the greatest of my teachers," said Kepler. The best teachers, however, do no more than point; it is then up to their students to find the way forward. An analogy nudges us toward an insight when close enough to conform to our problem and distant enough to transform it.

Mathematics is fundamentally the science of self-evident things.—Felix Klein

Descartes institutionalized one very broad sort of analogy when he invented coordinate geometry, which lets us translate back and forth between an algebraic and a geometric language; and this sort of mapping from one point of view to another is a deep part of modern mathematics, enshrined in category theory. But analogy often works less mechanically, reinforcing our old superstitions about inspiration, divine intervention, or the Familiar perched on your shoulder. We had an example of it before, where a way of proceeding in the problem of bisecting a pancake suggested the same sort of nonconstructive approach to a structurally similar but materially very different situation.

Let's look at an example where the unpredictable peculiarity, as well as intricacy, of analogy's working stands out. You want to know what the chances are of getting three heads when you toss a fair coin six times. The mind may dart at once from that three and six to "one chance in two"—an analogy reinforced by the argument that you'd expect heads half the time from random flips, and a second argument that you'll either get three heads or you won't, so again the odds would be one in two. The trouble with analogies is that they wobble: we can't be sure, at first, whether the way they fall short of identity will help or hinder.

Hold off analogizing further, and just look and calculate. The first coin you flip may come up heads or tails (H or T)—let's say H, and take care of the T case later. The second one again will be H or T—so we have two paths of a branching diagram,

Now we see the way forward: from each of these two branches two further twigs will sprout, and then two from each of them, and so on, making six ramifications in all: And let's not forget, the first flip might have been T, so exactly the same diagram follows from that. After six flips, you arrive at 64 possible H or T destinations, by 64 pathways (64, by the way, $= 2^6$, where the base tells you the number of possibilities at each juncture, and the exponent how many junctures there are). Now check out the pathways, and you'll find that of the 64, exactly 20 pathways contain 3 Hs. The chances, then, are 20/64, or 5/16: somewhat less than the innocent expectation of 1/2.

This result may be faintly surprising, but the method was sure, if tedious. It leaned, ever so little, on an analogy from events in time to paths in space: not a radical enough change in vision to strike us as due to a private muse.

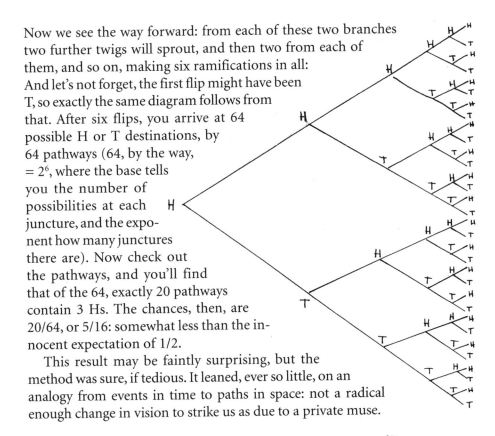

But one can make a slightly more farfetched analogy, which though less visually self-evident than the branching tree, leads to a numerical formulation of what is going on—one which generalizes more easily to more coins and more flips than you could or would draw trees for. What if someone now said that you could have avoided all that work, and saved some paper as well, by rethinking the coin-tossing problem in terms of *choosing* three heads-up coins from among six coins laid out in a row under a piece of cloth. Why? How would it help to replace the darkness of future time by literal darkness? And talk about the wobbliness of analogies: we're supposed to picture the coins as having already been flipped, with three heads among them, yet at the same time as not exactly already flipped, since those heads could be anywhere in the row of six hidden from view.

Let's follow cautiously on. We're asked to reach under the cloth and choose three heads from the row of six coins, which (a further, minute, analogy from number to position) we'll think of as lying in six numbered slots.

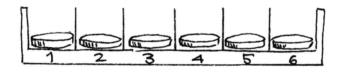

The first H we draw out might have been in any one of those six slots, so six possibilities for it. Having drawn that first H, the next H may lie in any one of the five remaining slots, and after it, the third H in any of the four slots left. That makes 6×5×4 = 120 possibilities in all for where the three Hs might have been. Notice, for what it's worth, a notational convenience: n! ("n factorial") is defined as *n* times all the counting numbers smaller than n. So 6! = 6×5×4×3×2×1; since 3! = 3×2×1, 6!/3! = 6×5×4.

All well and good, except we already know from the tree diagram that the answer should have been 20, not 120 (and anyway, 120 is far more than the 64 possibilities altogether). What went wrong? Ah; we allowed time to enter our imagining of space, and the problems of picturing time came with it. We chose a first H, *then* a second, *then* the third: say they had been in slots 1, 2, and 3. The result would have been the same had we first chosen the H in slot 2, then the one in slot 3, last the H in slot 1; or any of the six possible sequences (1×2×3 = 6, or 3!) of choosing the heads in three slots. Because the order of choosing doesn't matter to our result, we have to *divide* the 120 possible outcomes by 3!, giving us (6!/3!)/3!, or more simply, 6!/3!×3!, which gives us 20, as desired.

Remember that the first of those 3! came from rewriting 6×5×4 as 6!/ 3! Had we been trying to choose three heads from among seven coins, we would have had 7×6×5 possible arrangements to choose from, i.e., 7!/4!, or 7!/(7 − 3)!, and once again, we would have to divide by the 6 possible sequences: by 3!

In general, if you want to choose k things (in our case, heads) from among n objects (coins), there will be $\frac{n!}{(n-k)!k!}$ ways to do so. Statisticians tend to call this "n choose k", and write it $_nC_k$, or, more cryptically, $\binom{n}{k}$. *n choose k*

Thanks to an analogy that makes sense retrospectively, we now understand that the chances of getting three heads when flipping a coin six times is the same as the number of ways of choosing three heads from among six coins. Similarly, the number of ways of choosing five heads from nine coins, say, is $\frac{9!}{(9-5)!5!} = 126$, so the odds of getting five heads in nine tosses would be $126/2^9$, i.e., 126/512. After putting in the conceptual work, the calculation is far easier than making a diagram with 512 branches and launching our counting program.

Could it have been made easier still? What if the Angel of Analogy now whispers in your ear that the answers you seek already lie neatly laid out in Pascal's Triangle? How can this be—what have these numbers to do with choice or chance? You may recall that the nth row of Pascal's Triangle gives the coefficients in the expansion of $(a + b)^n$, if the first row is n = 0:

$$
\begin{array}{ccccccccccccc}
 & & & & & & 1 & & & & & & \\
 & & & & & 1 & & 1 & & & & & \\
 & & & & 1 & & 2 & & 1 & & & & \\
 & & & 1 & & 3 & & 3 & & 1 & & & \\
 & & 1 & & 4 & & 6 & & 4 & & 1 & & \\
 & 1 & & 5 & & 10 & & 10 & & 5 & & 1 & \\
1 & & 6 & & 15 & & 20 & & 15 & & 6 & & 1 \\
\end{array}
$$

*6 choose 3*

and so on. It also helps to call the first entry in each row the zero-th entry. And look: there in the sixth row, the third entry is 20: $_6C_3$ (you might check that the fifth entry in row nine is indeed 126).

It isn't enough to see that this is—or might be—so; we need to understand why. *We need to see the analogy.* Whoever first saw it (or each of us, on rediscovering it), might have reasoned this way:

If you flip a coin you will get a head or a tail: 1H or 1T. Flip two coins and the only possible outcomes are HH, HT, TH, TT. Since HT and TH are the same kind of outcome as far as we're concerned (one of each), we could rethink this as 1HH, 2mixed, 1TT. Just for convenience, choose HT as the way of writing "mixed", and signal HH by $H^2$, TT by $T^2$; the exponent simply plays the role it should: as in a×a = $a^2$, it is simply saying how many of them appear. So we have $1H^2\ 2HT\ 1T^2$. This is enough

*1 2 1*

like the second row of Pascal's Triangle to tempt us to insert a '+' between each pair of terms. We're encouraged to do so by thinking that the H and T we began with make up the whole world of coin flipping: $1H + 1T$ add up to the total number of possible outcomes. Two coins flipped (or one flipped twice) yield four possible outcomes, made up of $1 + 2 + 1$ kinds.

So with three coins there is only one way of getting all heads—$1H^3$—and one way of getting all tails—$1T^3$—but three ways of having two heads and a tail (HHT, HTH or THH), so $3H^2T$, and likewise $3HT^2$. And there you are: $1H^3 + 3H^2T + 3HT^2 + 1T^3$, or 1 3 3 1 for the eight possible combinations. The chances of getting two heads and a tail, for example, when flipping three coins (or one coin three times) are 3/8.

And is it sheer chance that the binomial coefficients, and the ways of choosing, and of calculating these odds, are all the same? Clearly not. Pascal's Triangle displays all the ways that two ancestors (call them a and b) can produce combinations of themselves when mated again and again: these are their offspring after successive generations—or, as we say, their *products*.

So if you ask how many ways there are to choose three of one kind (for instance, a) from among all the offspring, unto the sixth generation of the two parents (a and b), you are asking about $_6C_3$.

If you ask how many offspring (products) there are of a and b, in the sixth generation, which have three parts of a in their makeup, you're asking for the coefficient of $a^3b^3$.

And if you ask what the chances are of coming up with three Hs in six flips of a two-sided coin (H and T), you're asking again about the sixth generation, and those in it with three heads—so $_6C_3$, or 20—out of all the progeny, or products, thus far: namely, $2^6 = 64$. That makes 20/64, or 5/16.

Production, progeny, product lie behind these various appearances, showing themselves differently in each. This is what *really* is. Well, on reflection, this is a metaphor too, which has produced each of the others. What *really* is—what is being borne across into each of its avatars—must be something invisibly structural, not metaphorical at all, receding from each of its offspring. From our standpoint, analogy drops away like scaffolding once its work is done—which is why we take *its* products as real. Only when we too step back, from time to time, do we recall that the world we understand is inseparable from the world we make.

### Holding Hypotheses Like Birds

We seem constitutionally incapable of looking around us without trying to make sense of what we see, and the grander the scheme, the better. Give children any sequence of numbers and they will find pattern after pattern in it, even if those patterns aren't there. We can't but make hypotheses; the trouble comes in trying to shake ourselves loose from their grasp (think how hard it is to talk readers of the daily astrology column out of their belief, especially when the predictions are—like enticing analogies—just vague enough to fit both our current and wished-for circumstances).

What a mathematician needs to develop, then, is the right sort of attitude toward his tentative convictions. That "if" with which they begin should act as an open parenthesis, closing only with a proof (and then—again like analogies—the parentheses can drop away and the insight become part of your outlook). Fencing-masters since the Renaissance have advised their students to hold the foil as if it were a bird: tightly enough so that it won't get away, not so tightly as to crush it.

Here is a nice example of the grip we and hypotheses can have on each other. Put two points anywhere on a circle's circumference and draw the chord connecting them. Into how many regions has the circle's interior been divided? Clearly two.

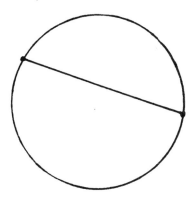

And with three points? Obviously four.

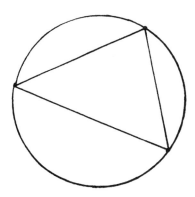

Four points give eight regions, when we draw all possible chords,

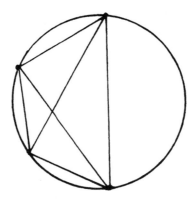

and five yield sixteen:

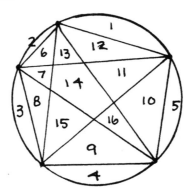

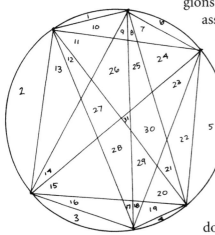

Anyone who hasn't conjectured by now that n points yield $2^{n-1}$ regions has been reading too quickly. Just to make assurance doubly sure, go back to the n = 1 case: does one point on the circumference produce $2^{1-1} = 1$ region? Yes.

Having seen five examples, the time is past to test—but ripe to prove. Oddly enough, this isn't as easy as it should be: why the number of regions should double with each successive point on the boundary isn't clear. Eventually you doodle your way to another example, for want of anything better to try. Perhaps when n = 6 the cause behind the doubling will become clear. For n = 6 we get thirty-one regions—which must simply be an error in our drawing—too many lines in too small a space; Wittgenstein was probably right when Bertrand Russell asked him why his *Tractatus* had no examples. "What," he is supposed to have answered, "dirty it up with examples?"

When we posed this problem to a Math Circle class of college students, one even came up with just thirty regions, because she had drawn a regular hexagon to start with, and its regularity caused a region to shrink to an internal crossing-point.

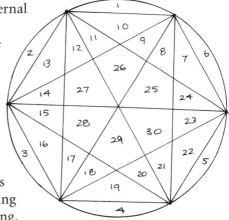

This must be the answer then ("Save the hypothesis!"). The question should really have been put this way: what is the *maximum* number of regions that all possible chords among *n* boundary points create in a circle's interior? Our thirty-one regions must have come from the six points not having been distributed maximally enough. As a friend of ours once remarked, nothing in mathematics is as difficult as counting.

For some reason or other, however, none of us were able to come up with the desired thirty-two, no matter how we arranged our six points. Two students got hold of a large piece of drawing paper and decided to see what would happen with seven points on their circle's boundary, while the rest—scornful of such pedestrian checking of the obvious—framed sub-hypotheses about interior points as disguising minute regions (think back to the line in the figure on page 31 that hid an area), or what a "really random" arrangement of boundary points would mean. Meanwhile, the students with their seven points could come up with no more than fifty-seven regions instead of the predicted sixty-four.

Facts dented our hypothesis but failed to break it. Our belief in order is too important for our physical and mental survival. Remember how long the Ptolemaic astronomers added epicycles onto epicycles, to keep some sort of fit between data and theory.

The students regretfully gave up their beloved powers of two and began to look at how boundary points, and the chords among them, create regions. Forced back and back, they had eventually to ask what a chord *was* and how it arose—and this clearing of vision proved to be decisive. Any two points on the boundary produce a chord; or differently put, a chord is, in effect, a *pair of points*. How many chords will n points therefore produce? One for every two points, so n points taken two at a time— oh!—$_nC_2$.   n Choose 2

That doesn't give all the regions, however, since many new ones are created where two chords intersect. And what does that mean? A new region for every *pair of chords*—in other words, one for every *four* points (two per chord): $_nC_4$. The total number of regions made by n boundary points should therefore be $_nC_2 + _nC_4$.

Try it. For n = 3, for example, $_3C_2 + _3C_4 = 3$ (since $_3C_2 = 3!/2!$, and $_3C_4$ doesn't mean anything, so adds no regions), but this is one less than the needed 4. We need a last bit of fine tuning: there was, after all, the one region of the circle we began with, prior to drawing chords—so the number of regions made by all chords drawn among *n* boundary points should be

$$1 + _nC_2 + _nC_4.$$

For n = 5 this does indeed give us 31 regions, and for n = 6, 57. A plausible argument and six confirmatory examples. But is this proof enough? We are so delighted by our success that we are sure our new formulation is right: can the hypothesis, having been released, now soar? When Jim Tanton presented this problem to a group of eight- to ten-year-olds in The Math Circle, in 2001, "They decided," he writes, "to count everything: the number of circles C (always 1!), the number of dots along the boundary, the number of paths drawn L (lines), the number of intersections I, and the number of regions P (pieces). It soon appeared that C + L + I = P and 'CLIP Theory', as they called it, was born. They noted that the theory was true for very simple diagrams and seemed to remain true if an extra path were added to a diagram. They reasoned as follows—as soon as this line intersects a pre-existing path, two things occur: the count of intersections increases by one, and a region is split in two, increasing the count of pieces by one. The formula C + L + I = P remains balanced. This equation also remains valid when the path eventually returns to the circumference of the circle: the path is completed, thereby increasing the number of lines by one, which is balanced by the fact that a final region is split in two."

*This is very neat & simple.*

### Experimental Fervor

When Sophus Lie was asked to name the characteristic endowment of the mathematician, his answer was the following quaternion: *Phantasie Energie, Selbstvertrauen, Selbstkritik* (Imagination, Energy, Self-trust, Self-doubt).—C. J. Keyser

Mathematics isn't experimental in the way laboratory sciences are. Its conclusions aren't ever more fully confirmed by repeated trials under controlled conditions but are assured, once and for all, by a single unbroken (if baroque) chain of inferences from assumptions. Something very like experiment, however, appears—as you have just seen—in the formulation of what we want to prove. Just as mathematics precedes experiment, in providing the structure within which science operates, experiment precedes mathematics in our ferreting out, hypothesizing, and constructing.

In these thought experiments, many of the qualities we've already spoken of combine: the confident sort of courage expressed in "*Versuchen wir's!*"; the precision that holds all else carefully still while allowing only one idea at a time to vary; the varying itself, which we described as rotating the diamond; the ingenuity, which is always applied analogy. But there are other aspects of the data, and of the mathematician's relation to it, that combine with these to give a special cast to experimental fervor here.

As for the data, you need only have seen one triangle in your life to know triangularity—and to know at the same time that in fact you have never seen and never will see a triangle at all, which is only badly approximated by steel gable-ends or mechanical drawings. What a peculiar science this is, then, with no single instance of the structures it studies! The instances, the variety of triangles for example, are instances of *form*, and the fact that we then operate on them formally gives another odd cast to experiments in mathematics: in them you can no more distinguish the means from the material than you can the investigator's mind from what it studies. These mirrored involvements add to the recursive character of mathematics, which so many have remarked on.

"The One remains, the many change and pass"—yet what mathematical insight feeds on is, in a deep sense, plural. David Mumford pointed out, in "The Dawning of the Age of Stochasticity", that computers make available for our scrutiny huge collections of data we had never before been able to collect, so that we may now hope to see unifying patterns in their variety.

Mathematicians develop the traits that experimentalists everywhere have, but they speciate to fit their remote Galapagos. They tinker, and come up with elegant devices (the generalization, for example, of exponential notation to rationals, reals, and complex numbers). They cultivate a quiet yet alert eye, which especially focuses on the peculiar, so that their seeing stretches from the singular to the single. They sometimes won't admit to, but nevertheless relish, the art of embodying: making miniature models as aids to the imagination. These models are no more real, of course, than what they model: not balsa and tissue-paper flyers, but each is a conceptual apparatus for thought experiments. So to picture where a triangle's centroid must be, you might think of a metal triangle, with the metal then made molten and drawn off equally to the three vertices, and the edges between them reduced to weightless wires. Now the laws of the balance beam will come to your aid (and all the while, your eyes are closed).

You catch the whiff of this abstract experimentalism best when mathematicians run into each other. "What are you working on?" A single diagram is sketched out, and a flurry of symbols adjusted, erased, expanded. "Have you seen this nifty idea—" and people who may have in common only the language of their craft, follow, with reckoned abandon, the trains of each other's thoughts on a crowded blackboard.

## The Architectural Instinct

Gauss on how he came up with his theorems: *durch planmässiges Tattonieren* (through systematic feeling around).

The desire to build ourselves into the world—to make a safe haven in it—must be among the deepest we have. Birds make nests from linear fragments of non-linear nature, shaping them so as to accommodate their lives to new locales: the crotch of a tree that supports, the leaves that disguise, this fairing of a little world into the large. Yurt and skyscraper equally serve the need to blend structure to structure coherently.

As with all desires, this one grows more rarefied as it intensifies, until structure itself, no matter what embodies it (images, sounds, language), becomes the material we would build with. Mathematics is the natural limit of this desire; its statements are ambiguously about the mind and the world, its proofs intermediate between the two, with immense plasticity, variety of design, and inventiveness. Their artifice builds inner to outer nature.

It is this architectural instinct that drives mathematical thought, seeking to contrive sound foundations and then to rear elegant structures on them, so that we will be at home in the meaningful world. The passion for order, the delight in symmetrical balance and asymmetrical tension, the sense that the whole, which is greater than all its parts, is the whole *of* those parts—all these are expressions of this instinct.

If it is so deep, if it derives from the yet deeper nest-building urge, then why are we not all mathematicians from birth, unfazed by the many vicissitudes this instinct undergoes as we bear it about with us? Love and fear are the answers. Love of the irreducible particularity of things, which sees a focus on structure as diminishing the value of substance; and fear of the abyss of abstraction, toward which ever more structural thought seems to lure us.

What turns someone into a mathematician—and you can see this in every Math Circle class—is that structures come to acquire the quirky individuality, the breath, of *things*. Just as a particular cast of light will call up urgent stirrings in an artist, or an interval, struck just so, in a composer, so a configuration—as small as a vanishing point, as immense as the whole of projective geometry—prodded at, poured over, puzzled about, will become a vivid companion. To pick out an example of this architectural instinct at work is perverse, because everything in mathematics exemplifies it. Still, here is one.

In our course on the Pythagorean Theorem, students know almost at once that the first instance they see is—just an instance: this drawing stands for any right triangle with legs of three and four units, and hypotenuse of five, no matter what those units may be and how the triangle is oriented on the board. Then they discover that the relation works as well for a triangle with lengths 5, 12, and 13, and all others of that ilk (since $5^2 + 12^2 = 13^2$, $(5k)^2 + (12k)^2 = (13k)^2$). The awesome generality of this is made up for by an awakening sense of what this statement, of what any statement, is about: not the accidents of here and now, but how things hold.

When the boy begins to understand that the visible point is preceded by an invisible point, that the shortest distance between two points is conceived as a straight line before it is ever drawn with the pencil on the paper, he experiences a feeling of pride, of satisfaction. And justly so, for the fountain of all thought has been opened to him, the difference between the ideal and the real, *potentia et actu*, has become clear to him; henceforth the philosopher can reveal to him nothing new, as a geometrician he has discovered the basis of all thought.—Goethe

They come up with a proof, and the same edged excitement pulses again: so it wasn't just true for triangles with sides proportioned to 3, 4, and 5, or 5, 12, and 13, but for all right triangles: in Borneo, on the dark side of the moon, or in spiral nebulae we haven't even names for, rotating and aging out at the edge of imagination.

Their proof follows inevitably from what it set out to prove—until they see or devise another, very different one. This must mean that the Pythagorean Theorem is true, independent of them, yet how they *know* it to be true depends on their contrivings. The architectural instinct is *for* the world but *from* the mind. It delights in growing the connective tissue between them.

Those infinite numbers of 3-4-5 right triangles have long since become particulars for them, in this immensely general context. And this general context? They now learn that not just squares but any similar shapes on the three sides of a right triangle obey the Pythagorean principle: however bizarre those shapes may be, the areas of those on the legs add up to the area of that on the hypotenuse.

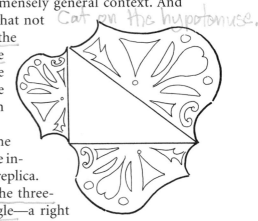

Our students are busy unpacking one of those Russian babushka dolls from the inside: each is contained in a yet greater replica. Then they unpack this latest one. In the three-dimensional analogue of a right triangle—a right

tetrahedron, where three of the faces are right triangles (the right angles all at a common vertex)—the squares

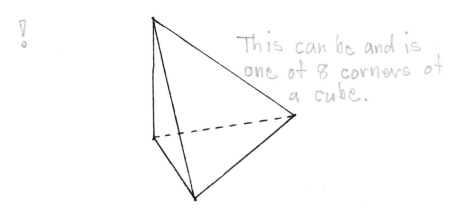

This can be and is one of 8 corners of a cube.

of the areas of those three faces turn out to add up to the square of the area of the fourth, "hypotenuse", face. They push further: this relation holds in the four-dimensional analogue too, and in the fifth, and sixth—in fact, in the $n$-dimensional "multi-orthogonal polytope", for any $n$, world without end (none of which can be drawn). And these are only some of the ways that the architectural instinct leads them from the universal theorem that Pythagoras came up with, 2,500 human years ago.

**The Conductor**

How could we possibly expect anyone—much less everyone—to have or develop the qualities we have been describing? Quite simply, because they are general human qualities, which can be adapted to the ecology of mathematics. None of us has them all, at an equal pitch, at any one time. They fade and flourish with the calls made on them, as do all our capacities.

The calls aren't separate, of course. We're not asked to be ornery and then, in a blink, fervently experimental. With the changing global and local character of the mathematical landscape we find ourselves in, these qualities (and doubtless many others) are linked up in coordinated sequences. By what? It would be most elegant not to have to invoke a detached Overseer—to let the calling and the ecology define the responsive structures in our minds. Think of this explanation, then, as a makeshift, an artifice devised by the architectural instinct to hold this aspect of inner and outer together.

Our orchestra of qualities has a conductor who brings out now this instrument, now that, in accordance with the interpreted score, that is, with a general strategy for coming to grips with the problem. Such a meta-quality may be an amalgam of all the others (like any conductor, who has to know the capacities of each instrument and player). Akin to the untensed attention we spoke of, the conductor looks beyond the immediate dispositions; he is concerned with matters of tact and timing.

What is curious, in the history of the doing and teaching of mathematics, is how often just this unifying quality is missing; a quality so likely sprung from common sense that you would expect it to be the most common. A decade ago two mathematicians even suggested that the practitioners of their craft be divided into those who speculate and those who prove, rather along the lines of experimental and theoretical physicists. While this view was vigorously attacked as "sanitized", condemning mathematics to "an arthritic old age" ("We now have high standards of proof to aim at," wrote Michael Atiyah, "but we must be prepared to act in more buccaneering style"), you will still find the stresses in mathematical thinking too often asymmetrically placed. Rigor is so late in development, and so essential for real mathematical thought, that in many a classroom it is taken to *be* mathematics, and spreads from its rightful place after invention to paralyze thought altogether. It would be just as distorting to let playful freedom, which devises conjectures, hang on to undermine their proofs, but we seem to err less frequently in that direction.

These aren't the only extremes that need mediating. When do you stop rotating the diamond, or following a deductive scent, to step back and look? When should you be as hard as ice, when yielding as water? We always need a regulator of our mental temperature. Would that the orchestra could perform by itself, as the Vienna Philharmonic once famously did; would that the music directed its own performance, even while—especially while—improvising. Perhaps this conductor is, at its best, the totality of the parts made into a presence among them: their modeled whole.

# five

# The Great Barrier Reef

I have yet to see any problem, however complicated, which, when you looked at it in the right way, did not become still more complicated.—Poul Anderson

By now you should feel assured that the vast continent of mathematics is yours for the exploring: nothing more nor less than being human qualifies you for the adventure. And yet, between our humanity and the abstraction of mathematics lies a Barrier Reef more formidable than Australia's. It is edged with an endless variety of off-putting symbols that, further in, harden into equations. Their desolate expanse is often drained of what little color it has by the impersonal exposition, which some people think suits universal truths. But say you make it past these outer defenses. The shore looks inviting—yet making your way inland proves far harder than you thought. The natives seem unfriendly; the interior is hardly visible—no wonder you succumb to Explorer's Malaise: that weary dismissal of the unreached as the unsought.

## Language: Symbols

$\sqrt{\phantom{x}}$, the square root sign, says it all. Angular, awkward, uncouth, it snarls from the page at us, daring us to come nearer. With so many whimsical and graceful forms among its antecedents,

The ill and unfit choice of words wonderfully obstructs the understanding.—Francis Bacon

$$\Gamma\ \mathcal{R}\ \sqrt{}\ \vee\ \mathcal{R}\cdot\sqrt[3]{}\ l\cdot\gamma\ \sqcup\ \sqrt[3]{}\ \mathit{v}\mathcal{b}\ \checkmark\ \mathcal{f}_3\ \sqrt[3]{}Q\ \tau a\ \Upsilon$$

has this one survived just to remind us of how off-putting mathematical symbols can be? In order to lead your thought on, they pull your eye up to exponents and down to subscripts, and even at times backward as well as forward, as in $_nC_k$. And it isn't just their shapes and the way they proliferate into a grotesque forest:

$$\iint_V (\omega_m - \omega) * (\bar{\omega}_n - \bar{\omega}) = \sum_{i=1}^{M} \iint_{V_i \cap V} \left\{ |p_m^{(i)} - p^{(i)}|^2 + |q_m^{(i)} - q^{(i)}|^2 \right\} e_i \, dx \, dy < \frac{1}{8^k}$$

not just *how* they stand, but *what* they stand for. You know it will cost you hours of deliberation to feel at home with just one of them, and by the time you've mastered the next, you'll have almost forgotten the implications of the first. Take square roots again: the other arithmetic operations you were brought up on turn pairs of numbers into a single number. This one produces two numbers for every one. Slip a small 3 into the crook of its elbow, and you have to absorb why the number its arm rests on has *three* different cube roots.

The intention wasn't to bring thought to a standstill but to hasten it on. Just as chess players, musicians, or meteorologists turn movements into squiggles so that anyone interested can read *through* them to the ideas they bear, so mathematicians reduce to no more than symbols the shunts that will carry a train of thought to the right station. How irritating it would be to have to read at length what you needed to know in and for an instant! How painful to repeat in full what you've already done a hundred times! And not just painful: you could actually lose your way in the tangle of words and fail to see the context with its hints of insight, from having to crawl so slowly from noun to verb.*

Symbolism is useful because it makes things difficult. Obviousness is always the enemy to correctness. Hence we must invent a new and difficult symbolism in which nothing is obvious.—Bertrand Russell  *Phooey!*

As an example, let's go back to the quadratic formula—but way back, to the ninth century, when al-Khwarizmi wrote his algebra text:

> What must be the square which, when increased by ten of its own roots, amounts to thirty-nine? The solution is this: you halve the number of roots, which in the present instance yields five. This you multiply by itself; the product is twenty-five. Add this to the thirty-nine; the sum is sixty-four. Now take the root of this which is eight, and subtract from it half the number of the roots, which is five; the remainder is three. This is the root of the square which you sought for.

---

*It is worth remarking that many mathematicians are as much put off by literary language as nonmathematicians are by math symbols. What uses are these turns and tropes, these allusions, indirections, and ambiguities, except to obscure the world and dim our sight? Each sort of language unfits us for the other, although the aim of the arts that employ them is the same.

Don't your fingers itch to translate this into symbols?

$$x^2 + 10x = 39, \text{ so } x = (10/2)^2 + 39 - 10/2,$$

or, as we now say:

$$x = \frac{-10 \pm \sqrt{10^2 - 4\,(-39)}}{2} = \frac{-10 \pm \sqrt{256}}{2} = \frac{-10 \pm 16}{2}$$

hence x = 3 (and also −13; we have come to allow negative solutions, since al-Khwarizmi's time).

The ten thousand symbols that fill up every crevice of mathematics are meant to act as ratchets do on a cog railway, keeping your thought from slipping back on the steeper ascents. When best used, they allow words to keep pace with thoughts, and each off-handedly stores an aperçu to delight in retrospectively. Each commemorates a victory of technique over the inertia of things. Why then do they so often grate on our nerves? For several reasons.

You're not angered and frustrated by picking up a newspaper printed in a script you don't know: you don't expect to be able to read it. But wodges of mathematical symbols occur in the midst of your own language, for all the world as if you were supposed to take them in as easily as words. You're wrong-footed even as you approach them. Practicing mathematicians are very familiar with this sort of thing, as when an overly general setting disguises the beauty of an insight ("This theorem," wrote the reviewer of a book on number theory, "is often in the context of a slightly stronger result, causing the simple and elegant ideas to get lost in mere notation.") If you ever had the misfortune of thinking you would learn mathematical logic from Russell and Whitehead's monumental *Principia Mathematica*, you know what he meant.

And yet the importance, if not the meaning, can glimmer through, as it did for the historian R. G. Collingwood when, as a child, he came on a book of Kant's.

> And as I began reading it, my small form wedged between the bookcase and the table, I was attacked by a strange succession of emotions. First came an intense excitement. I felt that things of the highest importance were being said about matters of the utmost urgency: things which at all costs I must understand. Then, with a wave of indignation, came the discovery that I could not understand them. Disgraceful to confess, here was a book whose words were English and whose sentences were grammatical, but whose meaning baffled me. Then, third and last, came the strangest emotion of all. I felt that the contents of this book, although I could not understand it, were somehow my business: a matter personal to myself, or rather to some future self of my own. . . . There came

*Here's aperçu for at least the 3rd or 4th time.*

upon me by degrees, after this, a sense of being burdened with a
task whose nature I could not define except by saying, "I must think."

In mathematics, the sheer unattractiveness of so many symbols doesn't
help. Gawky alone, contorted in clusters, it is almost as if the begetters
had no consideration for their audience. What seemed like a good idea
at the time often sweeps them away: $\cup$ might come from the $U$ of *Union*,
and $\cap$ looks enough like the $A$ of *And* to stand for intersection, while
$\supset$ looks like an arrow, $\Rightarrow$, of implication, so is a good candidate for *if-
then*, with $\subset$ for the inverse implication—so that (with ~ as negation)
we can turn a law of De Morgan's into a riddle out of Conan Doyle's
"The Dancing Men":

$$(\sim(p \cup q)) \supset ((\sim p) \cap (\sim q)).$$

It has to be said, though, that all those letters from the beginning of
the alphabet used as coefficients, and those from the end as variables,
with subscripts chosen from the middle, lets you make quite general
statements precisely and concisely, once the conventions are mastered.
There are many things to say, for example, about polynomials in general, that would suffer in the saying, were they to be said instead when
particularized to this one and that, or even this kind and that; so the
general polynomial looks like this:

$$f(x) = ax^n + bx^{n-1} + cx^{n-2} + \ldots + px + q,$$

or (so that you won't think they go on for at most seventeen terms),

$$f(x) = a_n x^n + a_{n-1} x^{n-1} + \ldots + a_1 x + a_0.$$

What's happened here? We've done away with the arbitrariness of a
twenty-six-letter alphabet by introducing subscripts that pair up with
the exponents, letting us talk easily about polynomials with any number
of terms (the notation alone suggests that we might go on to talk about
polynomials with infinitely many terms: a telling instance of language
paying back some of its debt to thought). You can see why you might
want to condense even this, at the cost of a few more conventions, to:

$$f(x) = \sum_{i=0}^{n} a_i x^i.$$

There's the glory and despair of symbolizing, in a nutshell. But as Oliver
Wendell Holmes once said, "We are mere operatives, empirics and ego-
tists, until we learn to think in letters instead of figures."

Another reason some symbols halt thought rather than help it, is the point of view they embody. Consider counting. "Take computation away from the world," said Isidore of Seville, some 1,500 years ago, "and all would be left in blind ignorance." But try computing with Roman rather than Arabic numerals, and you might soon prefer ignorance. Exponents, on the other hand, are a brilliant invention: they not only save mental effort but suggest, by their very presence, ways to extend their use: if $x^m$ means x multiplied by itself m times, so that $x^m/x^n$ would, after simplifying, just naturally be written as $x^{m-n}$, why might not the "number of times" be negative too, or some non-integral rational, or in fact any real—and then even complex—number?

All the while, of course, as we draft rules for manipulating these abstract symbols, those which were just as abstract to begin with, such as 1, 2, 3, come to seem not symbols at all but parts of the familiar world. Again, the responsibility for choosing what to symbolize, and how, lies more heavily on practitioners than they sometimes realize. We've heard that the physicist Wolfgang Pauli flew into a rage when a student of his, at the blackboard, chose to replace a long expression, which would need to be repeated several times, by a letter. "Your expression by itself makes no physical sense," exclaimed Pauli. "You have cut up reality wrongly. Don't you know that signs are not harmless?"

Symbols also go bad on us because people, traditions, and eras will each devise their own sign for the same idea, and then the generally poor sense that mathematicians have of history will let this multitude pour indiscriminately down on us. Think of as simple an operation as multiplication. Most of us learn addition first, so that multiplication comes as a shock to the system. But all right, you take the sign + and roll it over on its side, ×, so that 4 × 5 means what 5 + 5 + 5 + 5 used to (and oddly enough, is the same as 4 + 4 + 4 + 4 + 4: something to think about). You no sooner get used to what seems a streamlined sort of addition than you have to give up × for · (because x is about to be used for the dreaded unknown).

Never mind that the old multiplication sign was upper case and the unknown lower: this is one distinction too many for a hand hurrying through a calculation, and a mind struggling not only with tables that flicker in and out of memory but with the profoundly puzzling idea of a letter revealed as a number when put under sufficient pressure.

Two symbols, then, for the same operation (and a further complication in countries that use a raised dot as a decimal point), until the next turn of the screw: indicate multiplication just by juxtaposing two symbols! Thus, 5 × y became 5·y and is now plain 5y. Does that mean that 5·6 should now be written 56? Well—no. And should you write (x + 3) · (x − 4) or (x + 3)(x − 4)? That depends—on the traditions of your tribe.

Fine. You've been told that you must use dots for multiplying two numbers but *not* between a number and a letter. What about between two letters, then? Oh, that's just xy or yx, as you choose. All right—so xx? Er—that's written $x^2$. Where did that come from? And since 5·4 is the same as 4·5, and xy = yx, is $x^2$ the same as $2^x$? Actually, no.

You have to admire the wonderful flexibility of the mind, and its playfulness with language, that it can eventually master these shifting names for the same thing and the shifting rules of their usage. But then, we learn to read different type fonts, and worse, different handwritings: our capacity to abstract and generalize may be our most distinguishing characteristic, and the one on which our involvement with mathematics rests. Continue for yourself the secret history of the signs for multiplication, into n! and $_nC_k$, and beyond—and notice how what at first seemed streamlined addition reveals itself as an independent operation of growing subtlety.

The last way that symbols can mislead us comes, in fact, from this ability of ours to abstract. Things get out of hand so quickly in mathematics ("Do cats eat bats? Do bats eat cats?" asks Alice in Wonderland) that we irrationally pin our hopes on the symbols, trusting them to do our thinking for us. Let's go back to that Math Circle class of six-year-olds coming to grips with fractions. It was exhilarating to get halves and thirds in the right order on the number-line, and to add fractions all of the same sort (although a perfectly reasonable objection was raised against 3/4 + 2/4: how could 5/4 possibly mean anything, when 4/4 is the whole thing?) But 1/2 + 1/3 was another matter.

If you keep your suggestions to yourself at times like these, you will see Mind wrestling with the intractable world. The most popular surmise (inspired by the form rather than the meaning of the symbols) was that 1/2 + 1/3 = 2/5. The most puzzling—vigorously stated and defended— was that 1/2 + 1/3 = 4 1/2 (the reasoning turned out to be that 1 + 3 = 4, which took care of that annoying second fraction, and then there's 1/2 to add on). Other candidates included 2/3, 1/5, 1, and (after half an hour of wrestling) that there was no answer.

Symbols exercise an occult pull on the mind, especially when a wholly new idea (such as "common denominator") is needed to break their spell. No wonder we turn in despair to Kabbalah and childish incantation, hoping for clarity from a misty prism; aversion to thought leads to collusion with language.

If these are some of the reasons for our difficulties with the barrage of symbols in mathematics, what are some of the ways to overcome them? The best is certainly to invent your own symbols, keeping the power relation between you and them clear. In another Math Circle class for the very young, we were engrossed in counting up the number of roads in and out of the various cities on our increasingly complicated map (vertices and edges, to outsiders). We found ourselves faced with problems like "13 minus something is 8, what's that something?" and "11 plus something else is 19, so that something else is—what?" As their secretaries at the board, we grew tired of writing out "something" and "something else" and asked if they could come up with a symbol to stand for it (expecting "?" or perhaps "s", for "something"). Cameron raised his hand: "w", he suggested. "Good . . . 'w' for 'what'?" "No," he said, "w, because it looks like 'I don't know.'"

None of us ever had any trouble with $13 - w = 8$, or $11 + w = 19$, after that. Occasionally you need to mention, at the end of a course, that what we've called "tent-cities" are known as "points" to the rest of the world, and that "Anna's Theorem" is elsewhere called Euler's—but they take these failings of the world in stride.

Invented or come upon, a context of use grows up around symbols, which eventually moves them comfortably into the background of our thought. This context is at first of concepts, then of other symbols, as language takes on its role of place-holding, and its impetus thus gathers firmly behind you. Because students are eager to hurry on to using their symbols, they may be careless at first in making sufficiently expressive choices, and defining with enough precision; but conversation will tailor, and conjecture trim, to what works (and the admiration of others for a clarifying symbol further enhances the collegial undertaking).

A deep principle of our Math Circle is at work here: if you want students to master something—call it R—and R is a means to S, then work on S; R will slip unobtrusively in under the radar, whether it is a symbol, a technique, a lemma, theorem, theory, or point of view. This happens especially quickly with symbols, as testified to by the ease with which neologisms slip into and out of our speaking. The ripples of slang on the ocean of language delight the adventurer in us, as we roll with and cut across them. But don't sell short this development of familiarity through invention and use; once you win the battle with symbols, your agility in mathematics bounds sharply upward.

Learning how to make symbols transparent, so that we can see through them to the relations they stand for: this has been our theme. But symbols play a yet deeper role. At their best, they become visible again, but now as arrows that point to invisible structure. They aren't simply pictures of relations, but *stand-ins* for them: stand-ins we can manipulate when the structures are past our reach. This distinction between pictures and stand-ins underlies a deep division between the geometric and algebraic modes of mathematical thought. It isn't that diagrams are somehow undignified, or visual proofs inadequate, but we sense that mathematics is really about relationships, which may show up in the visible world but aren't fundamentally embodied in it. This is our architectural instinct at work. Even numbers aren't the ultimate referents; they are also pegs to hang relation and structure on. Ideally we want so to use symbols that moving them moves what they stand for, and makes them comprehensible to us. This is very different from the obscuring signs of number mysticism. It is a reaching of the mind toward structure itself, which is dazzlingly hard to see.

*Not true!*

In Raphael's painting *The School of Athens*, Aristotle wisely gestures to the world around him: this is what mind abstracts from, in order to see it with deeper understanding (since the lunge toward abstraction is never for its own sake, but to make the translucent transparent). Plato points up: that's where the light is. The last barrier to overcome in using symbols is learning to look not at but along them.

## Language: Equations

> Do not imagine that mathematics is hard and crabbed, and repulsive to common sense. It is merely the etherealization of common sense.—Lord Kelvin

If the symbols don't get you, the equations will. Most people's blood pressure spikes when an equation leaps out at them, because they assume they're supposed to understand it as directly as they understand sentences on the same page—and they don't. But equations are the punch lines you can't be expected to get until you've heard the story.

Take as simple and devastating a case as this formula:

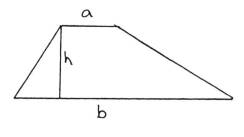

$$A_{\text{trap}} = \frac{(a+b)h}{2}$$

That is, the area of a trapezoid is half the sum of its bases times the height. It isn't much better when written out like that in words, since a reasonable reaction is, "Says who?" Says the teacher or the book, in a tone of finality, and you have one more "fact" to memorize unquestioned.

In a Math Circle class on the Pythagorean Theorem, we knew we would need this formula in order to re-create President Garfield's clever proof mentioned in chapter 4, but weren't about to thrust it on our students. Instead we asked them what the area of a rectangle was, and got the answer: base × height, or bh. *We* now asked *them* why, and they told us—after some discussion back and forth—that if you thought of area as a sum of little squares, then the rectangle was made up of b × h of them. And the area of a right triangle? They instantly saw that you could complete any right triangle to a rectangle with twice its area,

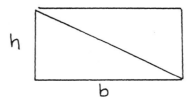

so the triangle's area would be bh/2.

What about the area of a parallelogram? This took more time, and eventually an insight: slice off a right triangle from one side of the parallelogram, slide it over and paste it onto the other side, and you've got

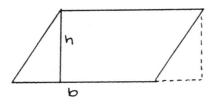

a rectangle again, with area bh, so that's the area of the parallelogram too. Well, so what's the area of any triangle, not just one with a right angle? And almost at once: any triangle is half of a parallelogram, just as a right triangle was half of a rectangle, so its area is also bh/2.

Now came what we were aiming at. What about the area of this awkward figure?

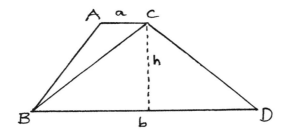

Since they had been engrossed in cutting up shapes and moving them around, drawing the diagonal BC was by now second nature to them—and there were two triangles, with bases a and b and the common height h—so the total area was ah/2 + bh/2, or—the punch line:

$$A_{trap} = \frac{(a+b)h}{2}$$

Equations come at the end of stories.

If you're not in a Math Circle class but reading a book and lurch suddenly into a formula or equation, what works well is to skirt it gingerly and let the context explain where it came from. The advice Roger Penrose gives (in his article on "The Rediscovery of Gravity" in Graham Farmelo's *It Must Be Beautiful*) is very good:

> If you find [equations] intimidating, then I recommend a procedure that I normally adopt myself when I come across such an offending line. This is, more or less, to ignore that line and skip over to the next line of actual text. Well, perhaps one should spare the equation a glance, and then press onwards. After a while, armed with new confidence, one may return to that neglected equation and try to pick out its salient features. The text itself should be helpful in telling us what is important about it and what can be safely ignored. If not, then do not be afraid to leave an equation behind altogether.

The formula we gave as an example hasn't the added dimension of fear that a full-blown equation has: namely, saying that two expressions, which look very different, are the same. A little equality sign is never innocent. It confronts more than two expressions: two lines of thought, really, giving the depth of a stereoptical view where before there was nothing to see. No wonder it takes some effort to look over, and then beyond it. Here is a monument to past work that ended with insight, saving you the trouble of reinventing *that* wheel, at least. And it is a pointer toward future discoveries, since its very form allows you to shift parts from one side to the other: adjustments that can bring into the open what was otherwise hidden (so in trying to find integers that satisfy the Pythagorean relation, $a^2 + b^2 = c^2$, nothing more than rewriting this as $b^2 = c^2 - a^2$ lets you factor the right-hand side into $(c + a)(c - a)$, and this—with evens and odds in mind—funnels you toward the answer.

An equation is, as Shakespeare said about poets, of imagination all compact—and that compactness, that condensation, is what scares us off. It is, say Ellen and Michael Kaplan in *Chances Are*:

> a kind of bouillon cube, the result of intense, progressive evaporation of a larger, diffuse mix of thought. Like the cube, it can be very useful—but it isn't intended to be consumed raw.

You're trying, after all, to put someone else's construction into your mind, and this takes both a kind of passive conforming of your thought to his (as an amoeba flows its shape and substance around what it would swallow), and an active, exhaustive, and often exhausting analysis that makes it, in the end, your own. You know you've mastered an equation when you no longer struggle to remember it. In fact, it then begins to do your remembering for you.

What makes the analysis of equations so trying is that since they are condensates, their parts, like the molecules in a crystal, vibrate within very narrow limits. No conversational leeway here, none of the fuzziness, of the more and the less, that lets us get on easily with people unlike us. Each term, each symbol, each relation is defined with a precision that takes getting used to, as builders of balsa-wood models learn to up the tolerance ante when they graduate to hardwood. Take an equation most of us feel comfortable with: in Δ ABC,

$$\angle A + \angle B + \angle C = 180°.$$

In words, as people in the middle ages would have encountered it:

If meas- The interior angles of a triangle are equal to two right angles.
uved

In the eleventh century the learned Ralph of Liége wrote to his friend, Reginbald of Cologne ("a man of powerful mind, who taught Latin to the barbarians of the Rhine"), that he had come on this sentence in a commentary of Boethius on Aristotle and had no idea what it meant. "Two right angles" is clear—but they could not be in a single triangle—and what were a triangle's "interior angles"? The very meaning of "are equal" becomes opaque. Both men were baffled. Reginbald eventually concluded that these were the angles formed when a line dropped from one of the angles met the opposite side. Once (he wrote), when passing through Chartres, he had revisited their old teacher, the distinguished Fulbert, and after many talks had convinced him that this view was correct.

The deepest cause of our awe, when faced with equations, is their very depth. The great laws of physics are conservation laws: they tell us that matter, or energy, or charge, or momentum, may never be created or destroyed, only transformed. Equations demonstrate the *conservation of structure:* what seemed immutably one thing is disguised as something quite different. And the beauty of these equations is that the disguises they deal in don't obscure, but reveal.

$\sqrt{2}$ , for example, is irrational, hence its decimal form must be chaotic: 1.41421356 . . . with no repeating pattern to those digits. But here is an equation that unveils a pattern for the $\sqrt{2}$ as simply intricate and infinitely extensible as anything decorating the Alhambra:

$$\sqrt{2} = 1 + \cfrac{1}{2 + \cfrac{1}{2 + \cfrac{1}{2 + \ldots}}}$$

An infinite series, which should be ungraspable in its immensity, is just playing a simple game, over and over.

But just because something can be described doesn't always mean it can be grasped: look, for example, at

$$1 - x + x^2 - x^3 + x^4 - x^5 + \ldots$$

Easy to see how this series is formed. But how possibly guess or even estimate what this endless sum will be for a given $x$, even when you're told that $x$ is between $-1$ and $1$?

But suppose the lights dim, and when they go up you see that this ungainly expression is simply equal to $1 / (1 + x)$:

$$1 - x + x^2 - x^3 + x^4 - x^5 + \ldots = \frac{1}{1+x}$$

as astonishing a transformation as any in pantomime. Yet there it is, an infinite series returned in a finite avatar—and to convince yourself of their equality, you need only multiply each by $1 + x$ (as long as x is between $-1$ and $1$: not an arbitrary restriction, but one which signals, as so often in mathematics, "Here be monsters.").

If an equation marks the moment when Mind unites in one expression a rabble of different forms, how much more impressive are those equations that summarize, describe, and predict natural phenomena. Think, for example, of the December 2004 tsunami. Its violent chaos can be modeled mathematically. "The resulting equations," says Gordon Woo, "yield a special combination of elastic, sound and gravity waves, which depict in mathematical symbols a scene of frightful terror."

Of the differential equation

$$mx'' + rx' + k^2x = 0,$$

Courant and Robbins say: "Many types of vibrating mechanical and electrical systems can be described by exactly this differential equation. Here we have a typical instance where an abstract mathematical formulation bares with one stroke the innermost structure of many apparently quite different and unconnected individual phenomena."

If these are the trials and rewards of equations "given", what about those we are obliged to invent ourselves? We usually encounter this obligation first in that peculiar literary form, The Word Problem. If you read

Two roads diverged in a yellow wood,

you're all at once there. But read instead:

A and B set out on two roads,
walking at 3 and 4 miles per hour respectively

and where are you? Suspended in abstract space with never a crossroads in sight or even a season of the year. What are you supposed to do now?

As the problem continues, with more information about when they left, and how far they were going, is some sort of relation among all these data supposed to dawn on you? Reason would piece it all together were it not stunned into silence by this barrage of numbers. You see why math teachers usually succumb to the temptation of telling you in advance that this is a "rate times time equals distance" problem; otherwise how would you know? It's enough to make you lose faith in form underlying appearance, not just because these apparitions are improbable stick-figure formalisms too, but because the order of understanding is all wrong: an equation like $d = rt$ has been told to you and is then imposed on reality, rather than showing through at the end like the ribs of rock that make sense of a landscape.

## Language: Third Person Remote

Mathematics is usually delivered to its audience with Spartan concision, Athenian elegance, and Olympian disdain. Intimidation follows and, more subtly, everything is made to seem on a par: the definition of a new term, the statement of an insight, its proof—so that you feel again that you are supposed to take it all in easily (when in fact years of effort may have gone into smoothing serious controversy into a phrase, or years of frustration into an innocent-seeming remark). The causes of this impersonal style lie in philosophy, psychology, sociology, pedagogy, linguistics, history, and even economics. The effect by and large is disheartening: gods tend to cut you no slack.

The economic reasons are straightforward. With thousands of intricate proofs published yearly in professional journals, space is at a premium, so the contents are freeze-dried. You are expected to rehydrate

HA!

them at your leisure, paying for the saving of public space with your private time. The constraints of journal publication reinforce deeper impulses to condensation.

The most benign of these is pedagogical. "It is left to the reader to show . . .", because the reader learns best by active engagement, and here is a push-up, a jumping-jack, to tone the muscles. It is when the reading turns into one of those Swedish training circuits, with new calisthenics to do every twenty paces, that you begin to wish you were back by the fire, being read to.

A kind of authorial impatience tends to creep in as well. There's the point you're eager to get to, and here is a half-imagined reader slowing you down with needless questions. As a ninth grade teacher once said to our son, "You don't see that? *I* see it!" And besides, you were leisurely, tolerant, and explicit at the beginning, when you spelled out your intentions and even defined the terms that needed no defining. Now you can pick up the tempo or, like the pilot of a giant jet, sharpen the angle of attack after takeoff. Those who should will keep pace: we're not in grade school now, as Hilbert said to a foot-dragging student.

> I recall once saying that when I had given the same lecture several times I couldn't help feeling that they really ought to know it by now.—J. E. Littlewood

Argh!

Some people undeniably relish dashing at top speed through their exposition (a leftover from whiz-kid days and calculating wonders) in the belief that brilliance is shown by taking the shortest way in these mountains: from peak to peak. A solid history is there to support them, back to Descartes, and before him, Aristotle. Descartes held that the cogency of your view was shown by the smoothness with which you expressed it (which is why our *ums* and *ers* become prolonged syllables in French). Aristotle said that wit consisted in detecting the missing middle term in syllogistic reasoning; twice witty, therefore, he who, with brevity in his soul, omits them for his audience to discover.

"*Il est aisé à voir,*" the eighteenth-century mathematician Pierre Simon de Laplace writes again and again in his *Mécanique céleste*: "It is easy to see."* His American translator, Nathaniel Bowditch—no slouch himself as a mathematician and astronomer—said that he never read this phrase "without feeling sure that I have hours of hard work before me to fill up the chasm." Impatience or pride on Laplace's part? What should we make of Laplace himself (according to his assistant, J. B. Biot) wrestling for an hour to understand what he had dismissed with an "easy to see?" Think of your IKEA filing cabinet finally assembled. Those plans lying in tatters

---

*Variants on this phrase include "It can be shown," "It is beyond the scope of this paper to show," and "This margin is too narrow to contain the proof."

of rage on the floor are self-evident, aren't they—*in retrospect?* And it is from this long view that math is written: now that you see it, how could anyone ever not have?

These winks and nods also serve to cement the fellowship in an exclusive club. But clubland plays another role in the genesis of a style both allusive and laconic. Published work grows in a conversational matrix, where ideas and language fitfully evolve. If you are in on this conversation you will have little trouble reading what amount to the polished minutes; if it took place in your absence, the minutes are from Mars. Even the art of writing minutes, however, can contribute to problems of understanding for the initiates. "No argument," says Alf van der Poorten, "tries to dot every 'i' and cross every 't'. Even correct arguments will contain gaps. But the trouble is that we may have conquered the apparent obstruction without realizing that we have now brought a real difficulty to light. It's that kind of thing that causes such admissions as 'my proof developed a gap, and that gap became a chasm that swallowed the proof.'"

*Mathemata mathematicis scribuntur:* Mathematics is written for mathematicians. —Nicolaus Copernicus

We've noticed, in some, an interesting side-effect of these difficulties. Writers of fiction and essays strive to say much in little, hoping the reader will be led by allusion to complete brief touches to a rich whole. We spoke of mathematical style as allusive and laconic too—but the allusions are of a very different sort (canonical rather than impressionistic) and the brevity comes not from understatement but portmanteau symbols. "Mathematicians have developed habits of communication that are often dysfunctional," writes William Thurston. "Most of the audience at an average colloquium talk

Does this book contain any abstract reasoning concerning quantity or number? No. Commit it then to the flames, for it can contain nothing but sophistry and illusion.—David Hume

*He's right.*

gets little value from it. Perhaps they are lost in the first five minutes, yet sit silently through the remaining fifty-five minutes. Or perhaps they quickly lose interest because the speaker plunges into technical details without presenting any reason to investigate them. At the end of the talk, the few mathematicians who are close to the field of the speaker ask a question or two to avoid embarrassment. . . . Outsiders are amazed at the phenomenon, but within the mathematical community, we dismiss it with shrugs."

Less savory are the uses of this way of speaking to keep clarity at bay. In 1895 the biologist W. F. R. Weldon wrote to Francis Galton about an attack on his conclusion by the statistician Karl Pearson: "Here, as always when he emerges from his cloud of mathematical symbols, Pearson seems to me to reason loosely, and not to take any care to understand his data. . . . Can we not get some mathematician on our committee? The position now seems to be that Pearson, whom I do not trust as a clear thinker when he writes without symbols, has to be trusted implicitly when he hides in a table of $\Gamma$–functions, whatever they may be."

The *Code Napoleon* has much to answer for here, as it does in so many aspects of life. For once you understand that law precedes practice— that there is a *correcte* way to deal with contingencies before there are any contingencies at all (there was a right way to make a *sauce béarnaise* before a cook ever set foot in a kitchen)—you understand too that form and the *correcte* are twins. Since the stuff of mathematics is form, it follows that it is the home as well of correctness, which is three parts elegance shaken with two parts élan. Oh yes, there's having insights and devising proofs—but as a mathematician we know was told in his student days, after proudly demonstrating a page-and-a-half-long proof, "Your proof is certainly not wrong. But elegance will come." These are some of the psychological components in the quality of distance that mathematical exposition features.

As the psychological turns to the pathological, we move from those who care that their audience can't follow, to those careless of it, to those past caring. Careless, out of an unexamined assumption that whoever wants to read on must be able to tell the strategic steps from the tactical (even though they are just numbered consecutively)—and, of course, math's ancestry in magic and mystery encourages a tendency to pull rabbits out of hats. Careless, also, because with natural or acquired autism they assume that their not understanding others means that others understand them perfectly well (a proposition not as irrational as it sounds at first, if you take it as a generalization of "you know what I mean," implying that since it is so obviously true, it would be embarrassing to spell it all out). Those past caring whether anyone follows are delivering an exposition to eternity—which brings us to the first serious defense of mathematics as presented in the third person remote. We will put this defense in the form of a little essay.

## The Buddha, the Bodhisattva, and the Bo

### The Buddha

Mathematics contemplates what *is*. None of the trimmings that are part of *becoming* therefore matter: who discovered its theorems, or how. Even the proofs devised by humans are no more than columns that raise up their architraves of truth into the still sky. If the proofs are to partake of Being too, it is only by stripping them of time's accidents and the idiosyncrasies of place.

How could we ever reach such conclusions as these? Instincts are the mind's equivalent of axioms, and interpretations of the world follow from them as surely as do theorems. From the architectural instinct, at any rate, derives this view of the world as form. It accounts for so much: the set theoretic language of mathematics, in which even functions are frozen to objects; the aesthetic, more Scandinavian than Victorian, that seeks out the spare, the blanched, the unupholstered; the medieval modesty of its craftsmen, who wouldn't spoil the lines of their constructions with maker's marks.

"Virgil shows me clearly what I fled from," Einstein once said, "when I sold myself body and soul to Science—the flight from the I and WE to the IT."

Having once glimpsed the *a priori* explains indifference to the synthetic. This is why mathematicians so notably lack curiosity about the historical, biographical, or even heuristic sides of their calling. This is why Gauss spoke for so many of his colleagues when he insisted on removing the last scrap of scaffolding from his monuments—freeing the Work from the trappings of work. This is why mathematics is typically praised by its practitioners for its chilling elegance, its impersonal attraction, its austere beauty.

The recursive character we spoke of earlier, the role of symbols as pointing beyond themselves, are at once symptoms, causes, and effects of this thinking about structure untainted by matter, this speaking in understatement untinged by irony.

"For a mathematical theory," said Arthur Cayley, "beauty can be perceived but not explained."

## The Bodhisattva

The Buddha view is from beyond the limit that converging series approach. With such retrospective looking it is, of course, "easy to see." Finding ourselves this side of those limits, however—uncertain which lines of our thought converge, and to what conclusions—we need any help we can get. If it is from those who have seen less darkly and have returned to instruct us, we shouldn't be surprised if they speak in tongues. Their language contracts as it draws away, and is meant to draw us with it.

When you approach an unfamiliar subject, you hope to work from whole to part, catching the shape of the thing so you can always sense where you are as you move into its details. But it is just this whole that mathematical style and language so often veil from our intuition, asking us instead to grant one thing and define another while creeping through deductive gates.

"Intuitive minds," Pascal wrote in 1660, at the beginning of his *Pensées*, "being accustomed to judge at a single glance, are so astonished when they are presented with propositions of which they understand nothing, and the way to which is through definitions and axioms so sterile, and which they are not accustomed to see thus in detail, that they are repelled and disheartened."

Still, a thoughtful mentor might reason that it is a different kind, a new way, of seeing that we must master, and one that can't be acquired gradually: you blink, you stare, your focus shortens and blurs—and all at once you simply see. There is nothing for it but to begin.

Think of the old joke about the child who comes home to tell his parents proudly what he learned in school: two apples plus five apples is seven apples. "Good," says his mother, "and how much is two bananas and five bananas?" "Oh," says the boy, "we haven't done bananas yet."

He hasn't a clue, you think: he hasn't *abstracted* to 2 and 5, from numbers as adjectives to numbers as nouns. When will it happen, and how, that he comes eventually to know (almost without knowing) that you can divide eight apples by 2, but not by two apples? That five apples minus two apples makes sense, and so (with a stretching of sense) does five apples minus six apples—but five apples minus 2 makes no sense at all? If you've forgotten how puzzling these abstractions can be, and how subtle their grammar, open a text in some branch of mathematics you don't know ("An **Eff** object (X, = $_x$) in Hyland's sense gives an assembly (A) with X as ambient, where each caucus $A_n$ contains all x such that n realizes x = $_x$x.")

A mentor who is sympathetic as well as thoughtful might let it be guessed that the structure built with this language has no shims: it is perfectly mitered, finished with emery paper. So the old cathedral builders, they say, fitted even the inner facings that would never be seen to the highest standards, for the greater glory of the structure and what it stood for—which, with the recursive cast of mathematics, might well be itself.

He might let on that language, left unadorned, becomes transparent: we are led to look through it—so that we see with new clarity the intuition we had abandoned. And he might so position you with respect to this remote vista that you came to see it as the supple vessel for your imagination, since its impersonality means that none is imposed on yours. What is almost the defining quality of mathematics—its recurrent generalizing that subsumes each latest breadth under a new expanse—follows from the very abstraction with which it is put. Mathematics is freedom.

## The Bo

The tree under which the Buddha found enlightenment was even more remote from human affairs than he became. Mathematics grows like a tree from its central core of number and shape, sending out ever finer ramifications that disappear in the higher air. We may scurry under its bark, or twitter awhile like birds in its branches, but it outlives us all in its grander rhythm. The beauty of this shape against the sky, and the beauty at its finest buddings are what we so love.

Yet for all that its branches are in the air, a tree has its roots in the ground, and those of mathematics spread out toward our common thought. We would delight in its forms no less were the access to them made easier. Is concision really the best way in? "One should strive above all else to avoid [an exaggerated desire for conciseness]," said the mathematician Liouville, "when treating the abstract and mysterious matters of pure algebra. Clarity is, indeed, all the more necessary when one tries to lead the reader farther from the beaten path and into wilder territory. As Descartes said, 'When transcendental questions are under discussion be transcendentally clear.'" Or as Poincaré put it: "Every time I tried to be concise I found that I became obscure, and decided I would rather be thought of as being a bit garrulous."

The ideal (beginning to be approached here and there on the Internet) would be to display mathematics fully, as a tree in cyberspace, so that you could find whatever you wanted to know about it and be led back, link by link, to the level that at last made intuitive sense. So the adult Hobbes saw a copy of Euclid lying open to the Pythagorean Theorem.

> He read the proposition. By G_, says he, (he would now and then sweare an emphaticall Oath by way of emphasis) *this is impossible!* So he reads the Demonstration of it, which referred him back to such a Proposition; which proposition he read. That referred him back to another, which he also read, and so on that at last he was demonstratively convinced of that truth. This made him in love with geometry.

This tree would not only have every statement that had ever been proven linked to each of its proofs, and each step in each one of them similarly proven, down and down (a vast Borges-like library in the only space large enough to house it), but conjectures, as they arise, like frail buds from each of the tree's extremes, along with false conjectures and mistaken proofs that had yet proven fruitful; and the whole continuum, from idle speculation to serious thoughts about these statements and about the growing tree itself, alighting and nesting and flying from it, moment by moment, back and forth to the neighboring trees of physics, and philosophy, and all the contending progeny of Mind.

We could range over this tree in whatever way our different styles of learning took us, lingering here, leaping there, zooming out or in, and then climbing higher—or out of that tree remote from us in cyberspace, and into its developing likeness in our minds, summed of personalities and now at last fused with ours.

## Climbing a Tall Building

In *The Revelations of Dr. Modesto,* a cult-novel of the late 1960s, Alan Harrington describes Mirko, the Human Fly. A man of invincible will, Mirko plans to walk in a straight line across America, no matter what heights or depths stand in his way. Special suction-cup shoes will let him walk up the tallest buildings.

What would Mirko have done had mathematics been one of them? For as William Thurston remarks, mathematics is a tall subject, with concepts built on previous concepts:

> It is possible to build conceptual structures [in it] which are at once very tall, very reliable, and extremely powerful. The structure is not like a tree, but more like a scaffolding, with many interconnected supports. Once the scaffolding is solidly in place, it is not hard to build it higher, but it is impossible to build a layer before previous layers are in place.

Will-power and suction cups aren't enough for scaling it, and the temptation to skip, or just touch and go, past the lower levels is very great. If you can slide past an equation, why not a chapter or a whole topic? The learner has no intuition of what is central, what parenthetical, what is structure and what illustration. There are no equivalents here of the toothpick bridges that young engineers delight in stress-testing; no toy or game resembles the upper reaches of mathematics even as closely as Monopoly does the arbitraging of adulthood.

Very well, Mirko shouldn't be walking up the outside and peering through the windows; he should be in where the action is. And once inside, won't there be express elevators? To paraphrase Euclid, there are no express elevators in mathematics. A little bit of darkness left behind— an ambiguity, a false assumption, a technique skimped, a concept misconstrued or overlooked—will rot the foundations out from under you.

You see it all in a wonderfully comic scene from Nicolas Philibert's film, *Etre et Avoir,* about life in a rural French school. A twelve-year-old is trying, with his mother's help, to multiply two numbers together.

"$3 \times 6$, 18, put down the 8, carry the 1—"

"Go on."

"I've finished."

"What? Isn't that a 5?"

"$5 \times 6$—25?"

"Recite the 5 times table."

Now the father sits in. "All right, I'm waiting. Where does it fit in?"

"There—no, there."

The father: "What's after the point?"

"Three figures."

"And before it?"

"Another three."

"Get your uncle to help you."

Soon the whole family—mother and father, older bother, uncle—are bending over the tortured piece of paper.

"How much is 6 x 2?"

"12."

"He's peeking in his book!"

"Don't forget to carry!"

" . . . plus 4—I'm lost."

"I'm lost too."

"There's a mistake there."

"He hasn't shifted it."

"You forgot this thingummy. It should go here."

"Here?"

"Yes. There's something wrong now."

"Where am I?"

"I don't get it."

"He has to shift it after. Otherwise—"

"What's up? 1 to carry . . ."

"12, and 1 to carry—13—"

"There's a mistake somewhere—"

"It's not right—"

"It's right—move your finger."

"OK, that's right!"

"No it's not! Look there! It's out by two figures!"

"No, that's normal!"

"No way!"

How could you possibly be fluent if you don't understand the idea behind the technique? And if you more or less understand it, the less will eventually shake the whole structure to pieces. We had a friend who managed to steer clear of the kitchen until his wife fell ill and he wanted chicken soup with matzo balls.

"All right," she called out him, "I'll tell you how to make it. First beat the egg-white into the matzo meal."

Half an hour later he came into the bedroom, exhausted. "I've been beating all this time and they won't mix together!" He had been trying to beat the papery skin inside the shell; it was, after all, white.

Many people, perhaps the majority of those who have run into math, have left it after the ground floor, where they first encounter signs for concepts and struggle to understand what it is that's being talked about: to understand that the apples were there only for the sake of how many there were; that numbers are things too, and somehow more long-lasting. Many of those who survive this will jump out of the second-floor windows, since it is here that the serious business of adding fractions is going on. Even the Greeks, our mathematical heroes, had trouble granting fractions the status of numbers. And now these have to be of the same sort in order to add them. If you come to terms with the abstraction of common denominators, *finding* them can still leave permanent scars on minds not yet nimble at division and multiplication.

But say you get through arithmetic with part of your ego intact. Each of us vividly remembers our awe of the big kids, who were up on the next floor doing something mysterious called algebra, and now we were supposed to be doing it too. Yet after all that work getting numbers and the signs for them straight, here was a sign thrown in among them that couldn't be pinned down at all. It didn't seem fair.

> Algebra: a peculiar science, which takes the quantity sought, whether it be a number or a line, as if it were known or granted, and then by the help of one or more quantities given, proceeds by undeniable consequences, till at length the quantity, at first only supposed to be known, is found to be equal to some quantity or quantities, which are certainly known, and therefore is likewise known.

This definition comes from a mid-eighteenth-century dictionary and expresses its author's perplexity along with ours. Little wonder that what immediately follows it is

> Algema: a pain, a sad troublesome sensation, impressed upon the brain from a smart vexatious irritation of the nerves.

We have barely come to terms with $x$ as the unknown quantity, when it transforms itself into a variable, taking on more shapes than the Old Man of the Sea. Many a child, confronted with $y = 3x + 2$, has plaintively asked, "Yes, but what *is* y?", only to be told that it depends on x. "But what is x?" "That depends." How can it depend? Does 5 depend on apples or oranges for how much *it* is? The issue is a deeper one: if you ask children whether $3·4 = 4·3$, they'll agree at once. "And does $5·7 = 7·5$?" Of course. "So ab = ba, for all a and b?" What? What are you talking about? Which a and b?

This is a classic example of how readily, once we master an abstraction, we incorporate it into our machine language and forget what it was not to know it.

These are the lowest floors in the endlessly tall, widening building of mathematics. You might think that having climbed successfully this far would mean that your skills with pick and piton would see you through to any height—but calculus lies one level farther on.

> Whoever finds himself lost in a labyrinth's complex of possible paths, never knows, from moment to moment, if he has made any progress.
> —Jean-Pierre Bayard

*Calculus*

*Yes*

People who rightly feel proud of mastering arithmetic and algebra give it all up when confronted with having to learn in a year what took more than two centuries to clarify: how to speak about change with fixed but ever diminishing numbers. We smoothly take in the slope of a straight line as the difference in a pair of its y-coordinates over the difference in the x-coordinates of the same two points—but "pair" means two *different* points. How can we be asked to find the slope as one point *becomes* the other? How can "approaches" turn into "at"? We are all as atheistic as Bishop Berkeley when it comes to the heathen gods Delta and Epsilon.

At each level in this building we think ourselves near the top—only to find that what seemed a skylight was an air-shaft. We tend not to notice until much later that we have become significantly more limber with abstractions, and our imaginations breathe more comfortably at these greater heights. We also begin to sense that the way up involves a way down: not only does the momentum for rising come from repeated recoils, but there are ramps of analogy here and back staircases of models there that let us return to the algebra or the geometry we knew and understand it better—and find that it also now helps us better to understand.

Mathematics has a final surprise for us: from time to time its building folds up like a telescope. How abstraction itself works—compacting the broadest structures we know to objects in structures broader still—begins to dawn on us, and we also catch glimpses of the building itself, reflected in its windows. What you understand on one floor subsumes the floors below.

All those differently inclined straight lines are species of a single equation; those lines, and the radically different shapes of circle, ellipse, parabola, and hyperbola, come from just slicing a cone at different angles, and those slices are themselves unified when seen projectively—and projective geometry lies in the gamut of geometries explained altogether by group theory, and . . .

This isn't a wholly benign process: a kind of amnesia sets in when all telescopes to nothing, and you forget the immense energy of imagination that got you there. Yet at these moments the slightest nudge can be enormously significant, lifting you up to the next floor. The chutes and ladders that materialize, as the building becomes surreal, work a change too on our seeing—our *structural* seeing—of the landscape it lies in: we see that it is all there, all of a piece; the connections belong to how we come to understand it. These paired but opposite insights differently satisfy our architectural instinct.

## Alienation

Redmond O'Hanlon describes this scene, deep in the rain forests of the Amazon: "Jarivanau and a small wiry man tried to re-measure their gifts, by stretching the fishing line diagonally from the big toe of the left leg to the forefinger of the right arm. The calculations, however, soon conjured up that look of blank pain and unfocussed anger that only mathematics can produce, and, with a sudden grin and a shrug, they gave up."

We're more than familiar by now with the nature, and many of the sources, of this pain and anger. What we want to focus on here is that *sudden grin*. For the pain would be pointless if, as in sports, it didn't lead to pleasures that couldn't be reached without the training. These are pleasures for all, no matter how little a person will later be involved with math: the pleasures of seeing the sense the world makes; of pride in your kind, for finding out how to uncover this sense; of heightened self-esteem, that you can master what seemed unattainable skills; and the pleasure of hearing this music without sounds, seeing this translucent architecture.

Those who are immersed in mathematics relish these pleasures daily—although this isn't the reputation mathematicians have in the world at large. Some see them as Oscar Wilde did, in his fairy-tale "The Happy Prince": "The Mathematical Master frowned and looked very severe, for he did not approve of children dreaming." A century on from Wilde's day, with more users of mathematics in the general population, this view has moderated, so that they seem less frightening than uncanny, like the patrons at the space bar in *Star Wars,* bleeping at each other in barely intelligible bytes. This view of mathematical culture puts off many who are otherwise attracted by its content (you may like Pictish designs, but would you have wanted to be a Pict?).

A few outlandish people, it's true, make their home in mathematics, and engrossment in its pleasures may short-circuit social banter; but while abstraction is a comfortable refuge for people oppressed by personality, geekishness isn't an entrance requirement. You'll find as wide a distribution of types here as in any community determined by interest: actors, sports fans, lawyers. What were the masons who built the pyramids like, and what did *they* talk about?

Now if in fact you can be of any sort, with regard to the rest of our human traits, and still prosper in mathematics—if the mathematical microcosm has well-filled niches corresponding to those of the world at large—then how has it managed to come by and to keep its reputation for being a club of obnoxious and arrogant children? Even some mathematicians seem to think of it so: Gödel reportedly said that "the gist of human genius is the longevity of one's youth," and the Russian Kolmogorov held that a mathematician's psychological development halts at the age that the math bug bites him (he thought of himself as forever twelve, but an eminent colleague of his as arrested at eight: the age at which boys pull the wings off flies).

Of course children needn't be obnoxious—this is taken to be a quality thrown in for free with the mathematical kit. It is beautifully illustrated both by the content and framing of an anecdote told about one mathematician by another: "Student X would declare in the middle of a talk: 'All this is trivial.' Taken out of situational context this may seem impolite, but actually this was quite productive. He pressed lecturers for more competence. Soon he had to emigrate."

We can certainly pinpoint one source of this unattractive side of mathematical culture. Just as some people think that children are fundamentally vulgar and need to have simple ideas explained loudly to them by goofy characters, others in charge of their upbringing believe that these little monsters will respond only to appeals to aggression, and—missing the point of mathematics as well—pervert the savoring of beautiful insights into opportunities to put one another down. *My* triumph (they get across) is all the greater for *your* defeat. This spirit, which feigns Greek origins but owes more to Leni Riefenstahl, has succeeded in giving what is meant to be a little-boy cast to much of the mathematical world (discouraging not only little girls but many an adult not charmed by the whinnies of success). It accounts for the otherwise anomalous practice of asking that problems be solved under time constraints and contributes, incidentally, one more reason for the in-language being cryptic.

But children are neither fundamentally vulgar nor fundamentally aggressive; they are fundamentally adaptable, and eager to take on the coloring of their surroundings. It is up to their mentors (as we suggested in chapter 3) to make the context of mathematical thought as supple as it is subtle, with the focus on the developing structure rather than on one's own prowess.

You would think this was especially easy in mathematics, which —by its emphasis on the architectural instinct—will draw to it those who are more comfortable with relations per se than relations among people, and here indeed may lie its attraction for many a preadolescent boy. With mathematical culture as it is presently constituted, this shyness blossoms in those joke categories of nerd, wonk, dweeb, and dork.

We once overheard some young mathematicians intent on getting the definitions of these terms right, as a mathematician should. To an outsider their conversation would have sounded self-ironic, to an insider, self-referential—what mattered was that in either case, "self" became satisfyingly remote. One of the more canonical definitions given was that a *nerd* was anyone who would engage in such a conversation. A *dweeb* was defined as someone who owned an automated tie-rack ("What's a tie?" asked a *wonk*; "What's 'automated'?" asked a *dork*). Some held that nerds flattened and nasalized their vowels, as a sort of shibboleth; others, that being a nerd meant you weren't aware of intonations, much less shibboleths. They concluded that in math, nerds can't help but get things right, dorks can't help but get them wrong, and dweebs aren't aware of a difference between right and wrong. This made wonks (said a nerd) the excluded middle.

That conversation gives the flavor of many a math camp and math common room, when problems and conjectures aren't being traded. How does its lightly self-deprecating tone square with the self-aggrandizement that camouflages the variety of personalities engaged in mathematics? Via the *innocence* of the egotism such competition brings to the fore. It isn't that a gene leading to social grace vies with one begetting mathematical agility for the same site on a chromosome, but that the abstractness of mathematics comes to infect what it touches, producing a kind of secondary autism. You see it in the unfocused smile of triumph, so like those on the faces of young chess champions: "It was a beautiful win!" A very shrewd friend of ours, the responsible daughter of a highly dysfunctional family, noticed this. "I wish I could be autistic, like a mathematician," she once said, "so that I'd no longer feel the strain of caring for my relations."

Is this, then, a community you'd like to join? Its ills are as superficial as those of any pursuit in its preadolescence. As more people come to dip and then plunge into math, their diversity will smooth away the oddities exaggerated by its small population now. The singular will take its place beside the general, selves will recognize and savor other selves. And when competition goes out of fashion, and collegial enjoyment replaces it, we will all look back on the spoiled children of these early days with the sort of affection that only distance allows.

And of course you hardly need sign up for life-long membership. Among the many subjects you study, mathematics—rightly approached—will promote a clarity of thinking, a playfulness of imagination and a freedom of invention that will enhance whatever you choose to turn your mind to.

Mathematics in gross, it is plain, are a grievance in natural philosophy, and with reason: for mathematical proofs, like diamonds, are hard as well as clear, and will be touched with nothing but strict reasoning. Mathematical proofs are out of the reach of topical arguments; and are not to be attacked by the equivocal use of words or declaration, that makes so great a part of other discourses, nay, even of controversies.—John Locke

## Diamond Hard

If math is our other native language, should we not speak it fluently, once all the barriers have been removed? Of course: as fluently as we speak the language we have been brought up with. Still, just as we need at times to search for the right word, the apt phrasing, we'll have to look now and then for how most elegantly to express a mathematical notion, or how to set up an equation or a deductive argument with the greatest transparency. Some prose, and certainly some poetry, takes careful reading to get the feel of the writer's style; and so does following someone else's proof. You don't read Shakespeare or Gauss with the TV on.

But just as there are thoughts too deep for words, no more than pointed to by language, so there are depths in the mathematical landscape that will take all our art to fathom. A gem of a mathematical idea, rotated the better to find an accessible facet, can be hard as a diamond.

The mathematician father of a young Math Circle student once told us he was sorry his son was enjoying the classes so much. Why? Because it might tempt him to become a mathematician, and that meant signing up for a difficult life. Although Pascal's father was also apparently discouraging, this parental complaint is very unusual, but anyone who has ever studied math, and everyone who practices it, has had hours, or days, or months when not superficial discouragements but intrinsic difficulties have made it out-top Everest, and plunge deeper than the Marianas Trench. A mistake in math is rarely trivial: it's not your facts but your way of thinking that you've gotten wrong.

Let's face this squarely. The difficulties begin with trying to learn a piece of established mathematics—in a course, or from reading a text or someone's paper. For all the feigned impersonality of exposition, the presenter's choice of style, order, and language is unlikely to be yours, and because the pretense is that the story is telling itself, you feel you have to rejig your mind to the way things are. While this isn't an intrinsic difficulty—not part of what's hard in the mathematics—it does lean a psychological weight on your learning: you are a supplicant at the shrine of eternal truth. And you must listen very carefully to the oracle. You have to gather where the stresses fall: here's an important theorem on the way to the target, such as the Pythagorean Theorem. Without it, we couldn't navigate through our world or build anything in it. And here's just a lemma needed to prove the theorem (one, perhaps, about congruent triangles). "Just a lemma": one step of many, so it shouldn't be high. Serious problems, however, can lodge in a lemma as well as in the theorem it steps up to.

What sort of difficulties are these? They almost always arise from the problem not being sufficiently in focus (are we talking about geometry or algebra?) and its context not being vivid or accessible enough. This is only made worse by not realizing it. The conditions of a statement, for example, may be subtler, or more (or less) restrictive, than you credited them with (the squares on the sides can face inward as well as outward); or you may have lost track of some (one of the angles has to be a right angle); or they may have consequences or antecedents that should immediately have come to your mind (entering, we often rush past the keepers at the gate with no more than the briefest glance: does this hold in higher dimensions, or generalize to triangles with no right angles?). A careful writer of fiction will choose his words to resonate with others in passages he wishes you to recall when reading this one; mathematical reference isn't oblique, but the symbol, the equality, the argument before you is to be read with its past occurrences in mind. So too inference plays the role of ellipsis: parts of a chain may be left out if the author expects his reader to fill them in from familiarity.

What we could call the romantic style of exposition will leave it up to the reader to know where he is and where he's going or—in the spirit of adventure—not to worry about it. This can tax your patience and powers of concentration, as you try to sort out tactic from strategy, and to understand fully where you are while knowing that it isn't yet where you wish to be. Losing the way can quickly lose you the end: "Who cares?" lies in wait down every blocked alley.

*Amen!*

The nature of language requires us to present arguments linearly, even when they aren't. We spoke before of the awkward symbols that make the eye dart around when it wants most to move forward; this is the complementary problem. Here you need to move around in the problem's context, seeing how the statement fits in to that which you do and don't know, to remote foundations, and to far-flung implications. This is what's meant by "getting your mind around it." And there's the exposition, marching steadily on. It takes reading and rereading, looking away, thinking sideways, seeing instead of saying, until the parts unexpectedly snap together and you've simply got it, and wonder what all the fuss was about.

A last difficulty in following someone else's mathematics—and one peculiar to the art—is the penchant everything in it has for leaping to the general. This leap often takes us where we want to go (seeing a problem that has filled our point of view now as a locale in a landscape that makes sense of its structure) —but we need to keep hold of where it and we are as the scale of everything slides about. "When we want to think about a mathematical situation," say Hilton, Holton, and Pedersen in their delightful *Mathematical Reflections*, "it is usually a good idea to think about examples of this situation which are particular but typical." Fetching from afar can bring you back, like Rip van Winkle, at a different age from your earthbound self, and the two need to be reconciled.

These are among the intrinsic difficulties in simply reading mathematics. They multiply when we set out to do the exploring ourselves. The situation really is very odd, when you stop to think about it. "We all find ourselves in a world we never made," says Sherman Stein in the preface to his *Mathematics: The Man-Made Universe.*

108

> Though we become used to the kitchen sink, we do not understand
> the atoms that compose it. The kitchen sink, like all the objects
> surrounding us, is a convenient abstraction. Mathematics, on the
> other hand, is completely the work of man. Each theorem, each
> proof, is the product of the human mind. In mathematics all the
> cards can be put on the table. In this sense, mathematics is con-
> crete, the world is abstract.

Even if you only take the invention and the proving as human work
(making congenial to our thought the way things independent of us
can't but be), why should our own inventing be so difficult to understand? For the same reason that self-analysis is notoriously hard.

To begin with, there's forming an idea of what the problem is: making it sufficiently suggestive without becoming impossibly vague. And
now what? Are there any strings hanging down from it to tug on? Our
intuition may not even incline us one way or the other about the idea's
likelihood (as with tiling that rectangle . . . or not). We steep ourselves in
the problem, we turn it around this way and that, we prepare the path,
but we always find ourselves speaking feebly, in a passive voice, about
what happens next—if it happens at all: the revelation appearing to your
prepared mind ("Come," sings the milkmaid to Lord Krishna, in E. M.
Forster's *Passage to India*, "come, come, come. But he neglects to come.")
This is the hardest part of what is hard about mathematics: having no
ultimate control over our insights. It is the price we pay for not being
automata, and the clearest evidence available that the order of the world
isn't quite the order of the mind.

Should you have your revelation, however, a problem of a wholly different order begins: proving it. You need to become a specialist in the art
of intermediate fictions. There are the axioms, propping up the vast palace; and here is the turret you hope to build out from its upper reaches.
You need to panel the interior and seal the surface, but above all shape
the struts that will carry this weight down to those foundations. The
enterprise has an almost linguistic character: you need to find expressions that will connect these remote extremes. Struts, expressions, connective tissue: what you hope to build are artifices that will allow our
collective way of seeing (summed up in those axioms) to look out of your
particular window. Think of an insight as a system of mirrors that, once
they reveal a pathway, can be replaced by a less angular proof. The number
of proofs that accumulate around the average insight testifies to how artificial in character, how ingenious in execution, this process of validating
is. And how precarious! Not a few proofs have turned out to have been
done just with mirrors, and needed to be revised or even abandoned.

You will find a new challenge when you step back from your theorem
to the theory it is now part of and try to understand how the context has

altered with this new content in its midst. Consequences reverberate backward and forward through the texture. If they unsettle more than they establish, you may have to rethink even what, by taking it for granted, allowed you to construct your proof. But this is the call. Answering it is the work of mathematics: a process almost indistinguishable from its product.

## Who Cares?

Go back to that telling passage from Redmond O'Hanlon: "The calculations, however, soon conjured up that look of blank pain and unfocussed anger that only mathematics can produce, and, with a sudden grin and a shrug, they gave up."

You see that shrug, you hear its vocal equivalent, "who cares?", in classroom after classroom the world over, day after day. What are we to make of Weil's remark about achieving knowledge and indifference at the same time? You could be pardoned for doubting its sincerity when you find Weil also writing: "As for my work, it is going so well . . . I am very pleased with it, . . . I am thrilled by the beauty of my theorems." We know this sort of ambivalence from Don Juan tales of amorous dalliance: no one matters more than Donna Elvira before the conquest; no one matters less after.

But something else is happening here: the peculiar rapidity, in mathematics, with which frontiers turn into suburbs. It is as if insights were just rollers that discovery trundled over. You see it most blatantly in what we could call "algebraic amnesia", since algebra indifferently stores up results without a trace of what led to them. *algebraic amnesia*

Take, for example, the stunning visual revelation that any square number is the sum of two triangular numbers. A square number, like $9 = 3^2$, looks like this:

```
. . .
. . .
. . .
```

while a triangular number is like a two-dimensional pile of cannon balls, with each row having one more dot than the row above:

```
              .              .
        .    . .          . . .
   .   . .  . . .      . . . .
. . . . . . . . . .  . . . . .
```

See Michael Leyton's "Shape as Memory" for
comments on this.
Math without memory

and counting up all the dots in each triangle gives us the triangular numbers:

1     3     6     10     15     21

How essentially different the two species appear—and yet looking at a square number slantwise shows that it is made up of two triangles. Wonderful! For its skew simplicity, this is an insight right up there with the best. *24 June 2007*

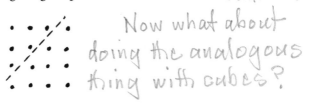

*Now what about doing the analogous thing with cubes?*

Now let's look at this relationship from a differently skew viewpoint: $n^2$, as we know, is a square number, but any triangular number can be represented as half the product of adjacent integers: take the triangle that represents 10, for example, push it over:

and complete the rectangle that contains it:

our 10 is revealed as half the 4 x 5 rectangle.

What about the next triangular number, 15? The same push-over-and-complete-the-rectangle maneuver will give us that 15 is half of a 5 x 6 rectangle.

Rephrasing this algebraically instead of visually and arithmetically, the two successive triangular numbers are:

$$\frac{(n-1)n}{2} \text{ and } \frac{n(n+1)}{2}$$

Add them, simplify, and mechanically find that of course

$$\frac{(n-1)n}{2} + \frac{n(n+1)}{2} = \frac{n^2 - n + n^2 + n}{2} = \frac{2n^2}{2} = n^2.$$

Where's the insight, where's the wonder, now? Who, you might rightly ask, cares?

The loss of meaning and flattening of effect, once a conjecture is incorporated into the formal body of mathematics, is yet another source, as well as an effect, of the "third person remote" style we spoke of earlier. Michael Atiyah called it "The Faustian Offer":

*You lose the insights ...*

> Algebra is the offer made by the devil to the mathematician. The devil says: "I will give you this powerful machine, and it will answer any question you like. All you need to do is give me your soul: give up geometry and you will have this marvelous machine." Nowadays you can think of it as a computer. When you pass over into algebraic calculation, essentially you stop thinking; you stop thinking geometrically, you stop thinking about the meaning. Fundamentally the purpose of algebra always was to produce a formula that one could put into a machine, turn a handle and get the answer. You took something that had a meaning; you converted it into a formula; and you got out the answer. In that process you do not need to think any more about what the different stages in the algebra correspond to in the geometry. You lose the insights. . . .

Most students of mathematics aren't afflicted by world-weariness after discovery. Their question is why should they care at all about what they've been set to master. Does the heart beat faster because 7 × 8 = 56 and not 54? Do the trigonometric laws rival the laws of attraction in shaping our destinies? While any subject can be presented boringly (focusing in a history course on when, rather than why, was the War of the Roses), mathematics is especially vulnerable, because means are mistaken by so many of its purveyors for ends (master these times tables; now these trigonometric formulas), and because the ways to make it life-enhancing rather than life-denying often seem locked away from them. You might almost think that a soulless people were out to induct the rest of us into their seedpod society.

The most pernicious form of indifference, however, surfaces, as it does in O'Hanlon's anecdote, after our best efforts are frustrated—and by something *merely* formal, *no more than* mechanical—something which, once you see it, will be dismissed with a disdainful "Of course!"

> Most fortunately it happens [wrote David Hume], that since reason is incapable of dispelling these clouds, nature herself suffices to that purpose, and cures me of this philosophical melancholy and delirium, either by relaxing this bent of mind, or by some avocation, and lively impression of my senses, which obliterate all these chimeras. I dine, I play a game of backgammon, I converse, and am merry with my friends, and when after three or fours hours' amusement, I would return to these speculations, they appear so cold, and strained, and ridiculous, that I cannot find it in my heart to enter into them any farther.

When you get as tangled up in mathematics as Jarivanau in his string, when it all seems as aimless as a labyrinth, one solution is to do as Jarivanau did—and give it up. But we're as likely to suppress our architectural instinct as we are to suppress our breathing. The other solution is to find again the thread that leads into the labyrinth and follow it to the center: follow it to *why* math matters, and why we cannot help but care.

We dismissed in chapter 3 the notion of a calling, but certainly some, for whom mathematics is irresistibly compelling, have thought that the compulsion came from beyond them. So Cantor, the father of set theory, wrote of a secret voice that spoke to and through him.

We know of only one instance where the call was recognized as such by someone who had never had the least involvement with mathematics. Kip Cooper, a young skier, had taken a bad fall on a mogul and ended up with a life-threatening spinal injury. As he lay in the hospital, paralyzed, heavily sedated before the dangerous operation, he had a vivid dream in which a voice demanded of him what he had ever done that could justify his life. "I was kind to children, I was kind to my friends," he answered. "And what about mathematics," asked the voice, "how much did you learn?" "That's not fair!" he answered in his dream, "what has mathematics to do with it? But if that's what you want, I'll climb the mountain of mathematics and teach it to children, so that they'll have a better time with it than I did, growing up." "On those terms," said the voice, "you can stay." When Kip recovered, he began learning the math he had neglected and is now, more than ten years later, in what he calls the dynamic loop of learning and teaching, each driving the other on. And every time he sets an equation equal to zero, it rings for him with overtones of his not having become zero, of his identity still preserved.

To speak of a calling is to shift the emphasis from mathematics to a broader spiritual realm, whose importance it inherits. How could someone who thus feels singled out for care, not care in return for what he is called to? Especially when (as many from ancient Greece to the present have thought), mathematics holds a privileged place in the making and knowing of the universe. "God ever geometrizes," said Plato; and Leibniz, two millennia later: "As God calculates, so the world is made."

This claim has always been most strikingly supported by a concord so pervasive as to be almost overlooked: the concord between mathematics and our understanding of the world around us. Why do we grasp the physics of the moon and the heart, of galaxies and ganglia, at all? Why do we understand them via equations, and why do these equations always turn out to be so simple—the more fundamental, the simpler?* Because of a "preestablished harmony" between physics and mathematics, said Leibniz. Hilbert and Minkowski echoed him in our own era. "Nothing takes place in the world whose meaning is not that of some maximum or minimum," wrote Euler: and calculus finds these extreme values for us.

Must we invoke a self more sentient than ours as the establisher of this harmony? Third person remote language gives us the impersonal shell of "nature" as an alternative:

> Our experience hitherto justifies us, [wrote Einstein] in believing that nature is the realization of the simplest conceivable mathematical ideas. I am convinced that we can discover by means of purely mathematical constructions the concepts and the laws connecting them with each other, which furnish the key to the understanding of natural phenomena. Experience may suggest the appropriate mathematical concepts, but they most certainly cannot be deduced from it. Experience remains, of course, the sole criterion of the physical utility of a mathematical construction. But the creative principle resides in mathematics. In a certain sense, therefore, I hold it true that pure thought can grasp reality, as the ancients dreamed.

Pantheism never wholly sidesteps theism. Einstein also spoke of a "cosmic religious feeling," which he said was the "strongest and noblest motive for scientific research." Dirac, one of the great founders of quantum mechanics (whom Heisenberg described as a "belligerent atheist") wrote:

> It seems to be one of the fundamental features of nature that fundamental physical laws are described in terms of a mathematical theory of great beauty and power, needing quite a high standard of mathematics for one to understand it. You may wonder: Why is nature constructed along these lines? One can only answer that our present knowledge seems to show that nature is so constructed. We simply have to accept it. One could perhaps describe the situation

---

*The most recent explanations (as of non-linear phenomena now) always seem arcane, and what they describe, fundamentally complex. If the past is any guide, these too will simplify in time, and as the enriched context of our thought accommodates them, will in turn reinforce our sense that understanding's ultimate surprise is its simplicity—not in the sense of a reductivist's empty formalism, but because complex phenomena in simple realms become simple in complex realms.

by saying that God is a mathematician of a very high order, and He used very advanced mathematics in constructing the universe.

The Nobel laureate physicist Eugene Wigner spoke, more simply, of "the unreasonable effectiveness of mathematics," saying that "the miracle of the appropriateness of the language of mathematics for the formulation of the laws of physics is a wonderful gift which we neither understand nor deserve."

The most extreme sect of purists, the Formalists, tried throughout the twentieth century to establish mathematics as a purely abstract system, like a game, in which the implications of a limited number of axioms could spin out an entire universe of relations. They drained mathematics of meaning in the hope of revealing its logical consistency. But even they write of the fit between physical theories and mathematical structures "as if this were the result of predetermination." At the extreme of this extreme, Dieudonné says:

> On foundations we believe in the reality of mathematics, but of course, when philosophers attack us with their paradoxes we rush to hide behind formalism and say "mathematics is just a combination of meaningless symbols. . . ." Finally, we are left in peace and go back to our mathematics and do it as we have always done, with the feeling each mathematician has that he is working with something real. This sensation is probably an illusion, but it is very convenient.

Our emotions, once we touch mathematics, are as undeniable as these explanations of them are suspect: for there is something of Lichtenberg's cat about them. "I have found people who were astonished," he wrote, "that cats have holes in their pelts just where the eyes are." A reading of Kant might suggest that the fit in the first case is no more astonishing than in the second: mathematics describes the way we make our observations thinkable, and physics is one of the ways we then knit these perceptions together.

This isn't at all to say that our reaction to mathematics reduces to "mere emotions," a threadbare irrationalism. These emotions themselves respond to the more deeply moving motions of the mind. For when our architectural instinct drives us to ferret out the patterns in our experience, along with the shapes we impose on it, we all at once find ourselves in the midst of timeless forms. Our experiences now with *these* are the stuff of mathematics, whose abstractions take on a secondary—then a primary—reality. We are on speaking terms with the infinite, and remote from daily care while caring passionately to understand the play of these forms, which in turn make sense of the world around and within us.

Who cares? We do.

# six

# How Math Has Been Taught

Was ever a human activity preached so differently from how it was practiced, taught so clumsily, learned so grudgingly, its light buried beneath so many bushels, as mathematics? It would take a separate book to count the ways this is so, much less to explain their intricate causes. Here we will survey the eccentric history of math teaching only enough to sketch a context within which The Math Circle rose.

Mathematics, as we have said several times already, responds to our architectural instinct—but different people, at different times, have understood this instinct differently and differently pictured how best to cultivate the response to it. This is why, from the Sumerians to the present, math teaching has swung like a pendulum between opposite extremes: between Mary and Martha—salvation by grace or by works: between math for itself, that is, and math as the servant of science; between math as formalism and math as intuition; between math as a body of discovered, infallible truths and math as an evolving product of human invention; between math as the material of thought and math as a way of thinking; between math for the millions and math for the elect—to name only a few of its polarities.

One constant in all this flux, which we'll take as a measure of how desperate the state of math teaching has variously been, is the approach that from the earliest written records has pinned down the bottom of the range: "If it looks like this, do that to it." Decorated with the elevated name of *Algorithm*, but commonly called Cookbook Math, it relieves the student from any need for thinking, and substitutes Truth by Authority for what could be dangerous encounters with reason. To find examples of it you need only open just about any door to a math classroom anywhere in the world, at any time.

This example goes back to Egypt and the Rhind Papyrus (dating from about 1800 BC). Its third problem asks you to divide 6 loaves among 10 men, and just gives the answer: 1/2 + 1/10 to each (with the exception of 2/3, the Egyptians only had fractions with numerator 1). But now you were supposed to check the answer, which you did by repeated doubling: 2(1/2 + 1/10) = 1 + 1/5. Doubling this, you get 2 + 2/5—but 2/5 isn't in your vocabulary, so you must resort to your handy "division table", which—without benefit of explanation—gave the results of dividing 2 by every odd integer from 3 to 101. Here you would have found that 2/5 = 1/3 + 1/15, so that you now have 2 + 1/3 + 1/15. Doubling this gives you 4 + 2/3 + 2/15—and again, the last fraction needs to be deciphered, so once more you unroll your division table to find that 2(1/15) = 1/10 + 1/30. Your grand total has become:

$$4 + 2/3 + 1/10 + 1/30.$$

You now look this up in your extensive addition table (!), and learn that the answer is 6—as desired.

Harder than seeing—as we would—that the answer is 6/10? Very much harder? Head-splittingly hard, and done in a cloud of unknowing? And yet this blind algorithm was used, around the Mediterranean, for two thousand years.

Should a student have asked why you do this, a common answer would be: "It works." The weight of tradition might have been added: "This is how it's always been done." Enthusiasts might, at a stroke, have replaced tradition by eternity: "This is how it is done." Should a student have asked *why* it works, few people would have understood the question, and it's unlikely anyone could have answered it.

The cookbook approach can't be pardoned, but it can be understood. In earlier days, when you didn't get to keep the book and your slate was wiped clean of one precept to make way for the next, you had to memorize. And if you were leaving school at thirteen to go into who knew what line of work, you'd better have been prepared for all, with a smattering of weights and measures, money exchange, gauging, and ready reckoning. It must have been like preparing young pilots for the Battle of Britain: hardly time to learn that if you do this the plane will dive, and if you do that the guns will fire—and no time at all for a course in aerodynamics.

But the cookbook approach also suited—and still suits—people who have gone into teaching math because it offers a quiet haven, where the lines are always neatly drawn and the equations beautifully lettered. Here are the instructions for preparing your homework folder that preface an old American geometry book:

> To the outside of the cover must be attached a label about 1½ by 2½ inches, filled out like the model below:

```
┌──────────────────────────────────────────┐
│                                            │
│   Winter Term              1902–1903       │
│                                            │
│              GEOMETRY                      │
│           Hamilton Fordyce                 │
│                                            │
│   Vol II        S&ND          I sec B      │
│                                            │
└──────────────────────────────────────────┘
```

Still

> This label may be neatly lettered in any style fancy may suggest but
> the order of entries must be the same as here shown. Before the first
> work is presented, attach the label to the cover, with its upper edge
> even with the leather back. Name and both city and home address
> must also be written or lettered neatly on the inside of the cover.

It is significant that the book presents the postulates of Euclidean geometry in exactly the same tone as these directions; they are equally valid, equally arbitrary. Do you catch, in the ritual purification of all this, a whiff of the obsessive-compulsive? These acolytes of the cleanliness that is next to godliness have cut out for themselves a semi-abstract space in the midst of life's chaos. The truths, the diagrams, even the lettering itself are impervious to calamity. This is how triangles must be in heaven, with vertex A always in the lower left-hand corner.

Romantics may look toward once and elsewhere for a Golden Age of Mathematical Tuition, when scholars studied patterns drawn in the sand and the stars, and lived in harmony with a thoughtfully cultivated nature. Where would this have been? In the Middle Ages, students began with the Trivium (not the three Rs, but grammar, rhetoric, and logic) and then graduated to the Quadrivium, where along with astronomy and music they studied arithmetic and geometry. Surely that would be a Golden Age curriculum. But much of arithmetic was spent in learning to cage numbers, as if they were wild animals, in ill-fitting cells (evens, for example, were led into the overlapping categories *pariter par*, or powers of 2; *pariter impar*, evens times 2; and *impariter par*, odds times 2). Most arithmetic, up to the fourteenth century, was taught by Bede's finger-reckoning. Division tended to be limited to certain numbers: Bede said you had to be able to divide by 59, in order to get the 29.5-day lunar month to fit the 59 half-day calendar (though in fact you could carry out this division using "Greek Tables", as he explained). By the sixteenth century, it was also important to know how to divide by 12, 20, and 28, for reasons of commerce.

Some problems are eternal. An answer from ancient Egypt: "Houses 7, Cats 49, Mice 343, Spelt 2401, Hekat 16,807, Total 19,607." What was the question? In nineteenth-century England, it had become:

As I was going to St. Ives,
I met a man with seven wives.
Every wife had seven sacks,
Every sack had seven cats,
Every cat had seven kits.
Kits, cats, sacks, wives,
How many were going to St. Ives?

By then, however, the answer had shrunk to One: only the speaker was going *to* St. Ives.

If you were avid for mathematics, where could you go? In the middle of the sixteenth century an Italian named Ferrerius taught mathematics in a Scottish monastery, using Boethius' Euclid for his advanced class. It contained statements of theorems—no proofs—from the first four books, then rules for measuring lengths, arcs, and volumes. He taught arithmetic from Sacrobosco's *De Arte Numerandi*: the rules were stated, no examples were worked out, and "the student was left to make a kirk or a mill o't," as a Scotsman, writing four hundred years later, put it. No wonder, then, that we read in the *Ludus Literarius* of 1612: "You shall have scholars almost ready to go to the University, who can yet hardly tell you the number of pages, sections, chapters and other divisions in their book to find what they should."

R. W. Southern, in *The Making of the Middle Ages*, says: "If a man wanted to study mathematics or logic he might have to wait for a chance encounter which sent him to a distant corner of Europe to do so: and in most cases, it is probable that the chance never came."

And out of Europe? Ancient Indian and Chinese mathematics were full of many correct and spectacular algorithms but without any indication of how they had been arrived at, or why they were correct, or what the limitations on them might be. Little chance, then, that they would have been taught other than as recipes. The great Islamic mathematicians were devoted to the Greek tradition of proof—but there is no indication that it was passed on in the schools. The art of reckoning on your fingers and on the abacus flourished among traders, which may have set a paradigm for the culture: the rules obviously worked, what need to inquire into their workings? Were curiosity aroused in this one or that, surely he would find his way to the House of Wisdom, or to a teacher who could satisfy his fervor. So among us, we learn to drive without needing to know how an internal combustion engine works, but *can* learn all the physics we need, if so inclined.

It is possible, of course, that somewhere in medieval or ancient times, East or West, mathematics was taught as a reasoned structure, founded on axioms and built up by logic, but that the texts have all since disappeared; or that something this important, or sacred, would not have been committed to writing (as the workshop secrets of other crafts were transmitted to apprentices by word of mouth). Possible ... but improbable enough for Occam's razor to shave such an apology away.

If the Golden Age doesn't lie in the past, perhaps it is just around the corner we've been careening toward ever since the Middle Ages. Again, we can take the prominence of cookbook texts as our guide. From the *Statutes of Charterhouse School*, 1612: "It shall be the Master's care to teach the scholars to cipher and cast an accompt, especially those that are less capable of learning and fittest to be put to a trade." Unlikely, then, that these skills were taught to the "less capable" other than by rote (evidence too of how far back the tradition of C. P. Snow's Two Cultures goes). And you certainly couldn't learn by reason what so much of school mathematics was devoted to: memorizing the different measures for different goods. Try keeping straight that while pounds and ounces were used for leather, tallow, soap, flour, bread, candles, resin, hemp, flax, and some Baltic goods, you needed lasts, sacks, weys, todds, stones, clones, and pounds for wool; pints, quarts, pottles, gallons, pecks, bushels, coombs, and quarters for grain, salt, sand, fruit, and oysters; and gallons, anchors, rumlets, and barrels for wine—but gills, munchkins, pints, quarts, and gallons for beer—to begin with.

The virtues of cookbook math were extolled in Thomas Dilwater's *The Schoolmaster's Assistant*, of 1743. This popular book went through forty-five editions in Great Britain, to 1816, and seventy in America, through 1832. Cast in question-and-answer form, Dilwater explains: "Children can better judge of the Force of an Answer than follow Reason through a Chain of Consequences." A book published at the time had proofs in it—and never reached a second edition. Another gave "reasons" in small print and suggested that they need not be read.

By 1800 textbooks were written for the teacher, not the pupil: "The text contains only such matter as is intended for the pupils to copy into their account books. By adopting this plan . . . teachers are enabled to reject the explanatory explanations without trouble."

A significant change comes with a growing realization—not that children might be capable of thinking, but—that their teachers might not be. By 1926 answers were printed in the back of the book. The 1939 *Beacon Arithmetic* was published with a full teachers' guide, with "reinforcement" for teachers "who might be unfamiliar with the material". It was used until 1955, although by that time the prices in its word problems were wildly out of line with the economics of the day.

It would be unthinkable now to issue a textbook without a separate book of answers, and a guide for teachers, and course plans, and chapter tests, and pedagogical supplements—not only because these satisfy an ever greater need, but because publishing is big business, with all the add-ons this entails. By the mid-1970s textbooks were produced primarily by foundations and publishers, rather than teachers or mathematicians. Some of

*Remember Miss Nelson and fractions*

*Yes!*

Standards are falling.—Cicero

these total packages were innovative, some traditional, with the consequence that a child changing schools often moved into an entirely different understanding of what mathematics is. Hence the clamor now, in reaction, for national curricula: a new Big Business.

Where do we stand? In 1966, 20,000 students earned math BAs in America: 3.8% of all BA degrees. By 2001 there were 11,000: and that was just .9%. The drop happens before students ever reach college: 11.4% of high-school graduates in 1975 were aiming for math and science in college; this had shrunk (despite the enormous new field of computer sciences, and despite the ongoing efforts to raise grades by dropping standards) to 4.4% in 2001.

In Great Britain the number of secondary school students taking A-level mathematics declined from 85,000 in 1989 to 54,000 in 2002, even though the courses had been restructured to be less demanding. Perhaps it will make Americans feel better to learn that (according to a 2005 report from the Manchester Institute for Mathematical Sciences)

> British mathematics postgraduates with a PhD from a British university are now largely unemployable in British universities. The level of research output, which British university departments are required to demonstrate in order to obtain adequate levels of funding . . . can now only be achieved by sucking in increasing numbers of older and more experienced researchers from overseas. Mathematics departments have no choice but to appoint the best applicants, and at present British applicants stand little chance of being shortlisted.

On the other hand, Americans will feel little better than the English when they look at the results of a comparison made in 2003 (through a straightforward test) among what were judged to be the mathematically most able fourteen-year-olds in the United States, England, Hungary, Italy, New Zealand, Singapore, and a few other countries. Fixing the average score at 500, an "advanced benchmark" was set at a score of 625. While 13% of all the students tested scored at or above this higher level, only 7% of American students did, and only 5% of those in England.

## Behind the Phenomena

Let's turn over this sorry tapestry and look, as mathematicians should, for its structural elements.

The teaching of math is always in crisis, thanks to what we might call, in an example of itself, "abstract recursion": the rising spiral in which the observed becomes a concept, which is then treated as an observation from which a new concept is derived. In math teaching, this means that as broader and broader generalizations come into view, those who see them try, mistakenly, to show them at once to those who don't, without providing the necessary intermediate steps that alone will make them meaningful. Having failed to get their

The study of mathematics is so easy that it affords no real discipline.—Sir William Rowan Hamilton

ideas across, those teachers then retreat to the research that engrossed them anyway, for which they are amply paid in self-esteem and the respect of their peers. Those who have soared won't want to waste any more time teaching fledglings to fly. Meanwhile, those teachers who don't see the greater abstractions won't know where math is now going, and so will teach it with outdated goals in view. Learning to divide by yet another number is no longer as exciting as it was in the sixteenth century and, in a climate that has itself grown more sophisticated, will seem pointless. And so two cultures arise *within* the field—research vs. teaching—with consequences as pernicious as those from the split between the sciences and the humanities.

As a result of all this, the means which teachers focus on go astray—or replace the ends. "First get the basics down pat" turns into "Practice makes perfect", and then "Practice *is* perfection". It's as if a teacher of English were to say "You'll love reading—here's a phone book!" Frustration at the failure to perfect or even to master these means leads to their fraying into a bizarre variety, which calls up sweeping reforms—that fray in their turn—and the all but organic cycle begins again.

You can see this as far back as ancient Greece, in the decay of elenchus to catechism. Plato's ends were deep and his means devious. One of them was to give to his dramatic character Socrates (a single, though long, remove from the historical Socrates) a method he called *elenchus*. This amounted to eliciting the truth from someone who didn't know it, by asking pertinent questions—on the assumption that he need only be woken to what he already knew from a prior existence. This method annoys Plato's readers (as he meant it to), because the truth seems not so much drawn from as thrust into the victim's mouth.

The classic example of elenchus is in the *Meno*, where an ignorant slave-boy is led to discover a line segment whose length is $\sqrt{2}$. Elenchus is certainly an ingredient of our approach to teaching math. We try harder to avoid ventriloquism, confining ourselves (more or less) to asking a fruitful question and trusting collegial conversation among our students to bring up the logical consequences and sort them out.

What is remarkable is the speed with which eliciting answers turns, in most settings, into a question-and-answer quiz. Elenchus takes too much time, too much patience on the part of the teacher, too much endurance on the part of the student. With a religious model in mind, and a sense of the immutability of mathematical truths, it is enticing to conclude that, as in the catechism, the ritually repeated question must elicit the ritually repeated answer. Answers are what matter, and these are in The Book, and teaching is preparation for the time when *you will be tested.* All of this coincides with the great shift from tutor to teacher: from what may become too intimate a relation between individuals to what edges away into an impersonal tension of one with many.

Because it is so fraught to teach your own children (so much emotion is invested in their success), sending them off to study with others has a long tradition. Monastery, Madrassah, and Yeshiva; the Masai huts where the secrets of adulthood are passed on by elders; putting your son, in the Middle Ages, as a page in some other noble's family—we take this unquestioningly in stride. Even the Hensley settlement of backwoods homesteaders, perched high in the Cumberland Gap a hundred years ago, wrote away for a schoolmarm from the valley to come up the broken track and live among them (is it significant that until modern times, girls tended to learn what they did within the family?).

Nowadays teaching usually takes place in the Classroom: that institution so common, for so long, among us that its peculiarity has yet to be fully appreciated. A teacher and students are thrown together—almost always by external forces—and must associate on unequal but intense terms for long volleys of brief encounters. Nothing in family life prepares us for this style of association that we have come to take for granted. The wholly different rules of engagement in office life afterward, or in the small societies of our choosing (sports, church, club) make us forget the bizarre conventions we once lived so fully by, and thought and longed and loved and hated within.

A student cohort develops along neolithic lines (scapegoats, heroes), with the alien Other of teacher putting in sudden and electrifying appearances. Many a teacher lives and dies for each least of them ("I was heartbroken by their final exams") and then—for the next year's. Teachers may be airily casual around the coffee machine in the lounge, since it wouldn't be professional to admit to others (hardly even to themselves) how much their hopes have become entangled with those in their passing care. These emotions rarely spill over into outright love or hate, but—standing *in loco parentis*—most teachers live in a haze of emotional perplexity, and most students catch enough of the confused signals to grow, intermittently, often reluctantly, confused themselves.

The components here span a peculiar terrain of human relation. Students may confide to a teacher what they can't bring themselves to tell anyone else, because the teacher is, in the end, a concerned but safe stranger (think of confessions made to gas station attendants). Teachers know that they will be measured by the success of their students, so their self-interest alone involves them in their students' work. In the course of the class, however, what they want is their students' admiration. At one extreme of temperament, they relish their dictatorship of a banana republic; at another, the handing on of a loved tradition and of the knowledge stored within it.

The explicit aim in a class—to master its subject—can wobble and fall away under the pressure of competing human needs: to get on with each other under these extraordinary circumstances; to establish and maintain a hierarchy and a way of speaking. A distance between teacher and student, casually called "professional", grows up to correct this wobble and re-establish the commerce of life, where we help others without having to invite them home. If artificial mannerisms develop, it is, after all, on the ball-bearings of manners that society glides forward.

That style of speaking soon begins to take over. Teachers learn to purse their lips and shoot laser rays of disapproval from gimlet eyes; they address invisible figures at the back of the room, and use "we" to mean "you". A silent contest of wills can decide who is called on, who ignored. Without any deliberate intention, many a teacher will neotenize his or her students to an age where rigid forms of response are accepted, and everyone has a nice day. The tone of exchange becomes more important than its content, and a class may end up without any class at all.

How does mathematics, in particular, fare in this constricted environment? Often very badly, since of all subjects, teachers feel least competent in math. Against a background of disdain for teachers in general ("those who can't do, teach"), those who are assigned—often against their inclination—to keep order in these least liked courses, even come to describe themselves as "sad old teachers" (rather than staying young with the young, they feel the age-gap widening away, as they catch fewer and fewer trendy references). They certainly can't afford to reveal their ignorance; access to hidden truth is their only source of power. Where knowledge is absent, form enters: $x$ is always the unknown, $a$ is always the leading coefficient, and the bar of the square-root sign must end with a hook over its right-hand-most content. Weighty issues are debated in the faculty lounge: should proofs be in flow-form or T-form? Should answers be boxed or underlined? In the sodality of the defeated, they may well console one another from time to time with the reflection that they have lived life at the chalkface, not in an ivory tower.

## The Teaching Wars

Crosswise to the perennial crisis, opposite tugs from the multiple polarities we listed at the start of this chapter deform the texture of teaching, locally and at large.

If math is the servant of science, then its study should begin and end with real world problems—posed from the start as messily as real world problems generally are, with extraneous data and ill-defined givens; and solved with error terms and approximations. But if math is derived from Understanding (the real meaning of *Mathesis*, as science is from Knowledge, *Scientia*) we must clear the paths to and within it of rubble, and let the mind enjoy the pure play of form. One of the earliest expressions of this opposition can be read into its Greek beginnings: is mathematics abstracted from the world (Aristotle), or is the world extracted from the forms that forever precede it (Plato)? Does form follow function, or function form? Schools and colleges that offer math courses designed—but never with quite enough content—to parallel the science courses taken simultaneously, fall in the first camp; those that stress the unapproachable quality of mathematical questions, methods, and standards of proof, the second. (While the French make this qualitative judgment explicit, with pure math at the top of their tree, it is implicit in most academic communities.) Inevitable dissatisfaction, in either case, periodically moves just about any institution—and at times the general fashion in math teaching—from one extreme to the other.

Proponents of mathematics as a formal enterprise have tended to hold the balance of power over the centuries, whether the forms were taken seriously in themselves or as a proxy for meaning. In epochs when geometry has been in the ascendant, intuition has overtly or covertly displaced formalism as the motive power: "seeing" a relation preceded its proof in value as well as in time. Worries about lack of rigor always swing the balance back to the formal and its spokesman, algebra; although geometric proofs can be fully as rigorous as proofs carried out in symbols, the latter bear a heavier impress of abstraction, and so read as somehow more sacred. But even when formalists are in the ascendant, they are always under attack from those who fear that their approach will desiccate mathematics by drying out the juices of insight. At the same time, new phenomena, in pure as well as applied contexts, urge us to find at least first approximations of explanation, even though these might lack the ultimate seal of rigor's approval. And just as in sports, the irrepressibly playful in us holds back from letting legalism bring invention to an end.

> Metaphor can draw the car of piety, but it tumbles the wagon of geometry into the ditch.—Augustus De Morgan

The "math is discovered"/"math is invented" dichotomy used to be innocent, the latter view naturally accompanying times of great mathematical ferment, the former, times of consolidation—and both leaving their traces in the teaching that scrambled to keep up. Now this opposition has been co-opted by rather boring political factions. Conservatives need an intellectual fixed point: God may no longer exist, but Math

126

is still omniscient. Even if you never understood it—perhaps especially if you never understood it—it has the comforting status of divine absolutism. If the social world is in terrifying flux, and morality has lost its bearings, at least Math is fixed and unambiguous.

The New Right of 1985 called what was then the New Math (investigation, problem-solving) "the latest stage in a disastrous process that has seen school mathematics drift toward becoming a low-level empirical science." The New Left sees in math an easy and early opportunity to make independent discoveries, untrammelled by the immense data-gathering needed in other fields—but then, ashamed of its elitist past, would put all mathematical ideas, however misguided, on a Politically Correct par, where sincerity displaces rigor as the touchstone, and such lurid flora as *Ethnomathematics* sprout. It is just common sense, of course, to take into account the context of life and thought in, for example, ancient India, when studying the rise of number theory there; or the Greek context, when looking at how geometry and deductive proof arose in its midst. The silliness starts with claims that mathematics follows from rather than underlies cultural phenomena; that rather than being universal it is fundamentally different in its Eastern and Western varieties, and speciates with national differences; and that one can only fully understand and enjoy the mathematics of one's physical rather than spiritual ancestors. Here too you will find that recently fashionable reinterpretation of Karl Popper's "falsifiability" criterion as a desirable end: mathematics should be in perpetual revolution, with any hint of its having unearthed a morsel of Being discarded as antiquarian thinking. So Right and Left misread as centrifugal or centripetal the rising spiral of abstract recursion.

Is mathematics the material of thought or a way of thinking? The first hardly differentiates it from the other sciences, taking it as "out there", to be discovered. Mathematics as a way of thinking received its greatest impetus from those Formalists who sought to *logicize* mathematics, giving logic and math alike the common language of set theory. They rode high, as far as teaching went, in the 1970s, with that New Math, which left parents unable to help their offspring decode Venn Diagrams, and teachers at a loss to decide which solution sets were complete (though it was rather fun rolling the Well-Formed-Formula dice, and filling in truth tables).

Which is the proper clientele of the mathematical enterprise? Lancelot Hogben's 1937 *Mathematics for the Million* wasn't the first attempt at popularization. For that you have to go back at least as far as AD 100 and the *Introductio Arithmetica* of Nichomachus of Gerasa, which entertained schoolchildren for a thousand years with its numbers dressed up as fanciful animals. And Cocker's *Arithmetic*, published in 1678, was so popular for the next century that "according to Cocker" entered the language as a synonym for "accurate." And why shouldn't everyone have access to this language, deep in the human brain?

But the view of mathematics as the province of the elect has always rivaled the view of it as a universal birthright. Its origins lie, perhaps, in the ancient world, where priests kept to themselves the knowledge of how the algorithms worked, whose husks alone they transmitted to what was itself a small group of scribes. Perhaps an inner circle of Pythagoreans, in Greece, knew the dark secret about the irrationality of $\sqrt{2}$; certainly what mathematics they did teach to their initiates wasn't meant to spread beyond them. You needed a credit in geometry to enter Plato's academy—almost as if it were a talisman. Old ways in olden times? It would be interesting to study these days the elitist hierarchies in Gifted and Talented groups, where the deep talk only to the profound, and the profound talk only to—themselves. Are there quiet networks of funding and coaching for young students identified as having genius potential? Are mathematical Olympiads nodes in such networks?

Attempts at makeshift syntheses of these opposite views are by no means recent. The 1868 report of the Taunton Commission, in England, concluded that "mathematics" should be taught to children of the upper-middle class and professional classes; "arithmetic and the rudiments of mathematics beyond arithmetic" to the mercantile class; "very good arithmetic" was for the sons of smaller tenant farmers, small tradesmen, and superior artisans; while working-class children were to get elementary arithmetic. We wince at this now—but we let the mathematical fate of people be sealed by their performance, in adolescence, on a three-hour multiple-choice exam.

The excessive study of mathematics is deleterious to the mind.—Sir William Rowan Hamilton

## Cookbooks, Song-Lines, and Games

We've gauged the lower reaches of math teaching by the extent to which it consisted in having students do no more than memorize recipes, and said that this dreary waste of mind and time has often been dressed in the fancier togs of "mastering algorithms". Perhaps this was a bit unfair, since algorithms condense great efforts of understanding into usable formulae, facilitating the next move upwards in abstraction. The mortal error comes in ignoring the meaning, history, and context of these efforts—skipping what, how, and why these algorithms facilitate—and simply presenting them as facts, no different from figures or from rules of symbolic manipulation. This stunts thought by demoting math to the level of incantatory magic, and accelerates the decline from shaping questions to memorizing answers. At its extreme, it takes the edge off the mind's finest capacity: to size up the unknown.

We can imagine two sorts of defense being made out for the cookbook approach to teaching. The first is that people desperately need a degree of awe in their lives: awe of authority (as represented by the hieratic figure of the teacher); but even more, of holy writ. Cultures with sacred texts often think it important to commit them to memory, or at least to attend their recital, whether one follows the words or not. What matters is being in the presence of The Word, in the language in which it was originally spoken. So with the Mass in Latin, the different parts of the liturgy resounding together in Orthodox churches, the spoken Talmud, the chanted Koran, or the Tudor English of the Book of Common Prayer. We have our literary talismans too: any Greek can intone the first few lines of the *Iliad* or *Odyssey* for you, any Russian the beginning of *Eugene Onegin*, any American the first sentence of the Declaration of Independence. Generations of English students say "To be or not to be, that is the question," and "Romeo, Romeo, wherefore art thou Romeo," even if they think "wherefore" means "where". The theory (if that isn't too strong a word) behind the vast majority of math teaching has been to dig it up from the page and plant it in the memory, where it may root, flourish, and propagate—though, more likely, rot and wither. Never mind: you will have been in The Presence, and therefore saved; you will pass the test, and therefore make your way in the world. Even better, you will know that your worth has been successfully measured against a platinum rod in the only Bureau of Standards that matters.

> Mathematics is no more the art of reckoning and computation than architecture is the art of making bricks or hewing wood, . . . or the science of anatomy the art of butchering. —C. J. Keyser

The other argument is secular. A good teacher knows the limits of a student's curiosity. The young love patterns but don't much care about explanations or justifications; watching them unfold and repeat, catching the lilt of their song, is enough. Why distract with trying to prove what slides so easily into memory? FOIL (First, Outside, Inside, Last) is all you know and all you need to know about binomial multiplication, SOHCAHTOA about trigonometric functions.

> Sally tarries by the gate:
> Four times seven's twenty eight.

That's from the 1839 *Marmaduke Multiply*, with its pretty woodcut illustrations. Sally and her gate may have no more relevance to four times seven making twenty-eight than they would to four times twelve making forty-eight, but picture, rhyme, and product enter the mind of childhood together. Older students have begun to gravitate toward what will be their lifelong interests, which might just possibly involve mathematics as a means—but hardly ever as an end. Why then punish the majority with proofs of what they need only apply? It suffices for most people that *someone* knows the reason why (this harkens back to the prior argument); the student's "I'll take your word for it" is the antiphonal response to the teacher's "'Shut up,' I explained."

These two defenses of teaching math by recipe stem from the world-weary conviction that people just can't *and shouldn't* be bothered with going back and back in explanation. Being educated means learning things, and "minus times minus is plus", like the rules for *which* and *that* can be explained, or made vivid, but the point is being able to get them right. Claims that stepping back better lets you leap forward, and that knowing the truth will set you free, bounce harmlessly off impatience, and our confidence that trial and error will see us through.

Upper-level math courses, from high school on, begin to leave the cookbook approach behind and teach math more as theory developed in a sequence of theorems, with the theorems presented as the conclusions of proofs. Now, you would think, Mind has come into its own. But no; the common practice has to be seen to be believed. The instructor (teacher, teaching assistant, graduate student, lecturer, professor) enters stage left, spends a few minutes on administrative details, then approaches the board and spends the rest of the class reciting definitions, theorems, and proofs—and writing them out. Everyone silently copies all this down. When a student once asked the instructor if she might photocopy and hand out the notes, he answered in puzzlement, "Then what would I do during the class?"

Some people may teach like that because they have just come from, and are just about to go back to, their research, and know the material so well that they can painlessly rattle it off in this way. Some are merely carrying on the tradition in which they were taught; reading is, after all, what "lecture" means. Others believe that the greater the variety of ways in which students can encounter these difficult ideas the better, and watching them unfold right there before their eyes has a dramatic force to it; and besides, questions will be addressed in problem sessions, or in conversations among the students themselves, afterward. A more serious justification is that the particular order of the material the instructor chooses, the particular proofs, the variations on definitions and lemmas, gives his particular take on the theory and a point of view that gets across in little asides, pointed emphases, and throwaway lines. This is how the touches of the workshop, the style of the master, and the secrets of the guild, are communicated.

Behind these various reasons, we suggest, lies a deeper explanation. You begin to detect it in the relish with which an instructor performs his work, writing with increasing brio as the class goes on. He is performing a piece of music. To know, to understand, a corner of mathematics is to love it, and to find ever new depths and implications in it each time you think it through. This writing up of the ideas is as much for his own enjoyment as for that of the audience, but it is crucially for the audience too: these are the song-lines of the tribe, the map sung into existence for each apprentice then to be able to follow on his own. Such, at least, is the fervent (if often unacknowledged) belief in the mathematical community.

It is a wonderful and inspiring undertaking—for the performer. But is it in fact the ideal way for the novice to learn his craft and the craftsman to become an artist? Put aside the niggling problem that should you lose for a moment the thread of exposition, there is hardly a hope of picking it up again: Theseus has disappeared around who knows how many turnings of the labyrinth. Put aside the cramped writing, the errors in notation, the sudden change in tempo from the hour's *adagio* beginning to its *accelerando* end. No matter how thoughtfully the definitions are phrased, nor how well the material is calligraphed nor how beautifully sequenced, the route that should have been discovered is no more than being retraced. The student will have to unlearn and reinvent it to make it his own.

Song-lines at one extreme, cookbooks in the vast middle—and what are the practices that fade into cookbook, or rival it, along the youngest margin? You'll find one attempt after another, especially recently, to lure children into math by making it fun. "Manipulables" replace memorizing times tables; a pattern is discovered, and then another; shapes are folded; bells rung; numbers dance. All this has the welcome effect of not giving fear and loathing even a look-in: games become the arena where minds encounter math. The problem usually is, however, that these encounters stay superficial—a decorative rather than an architectural instinct is catered to. No topic is dwelt on long enough to open up its depths, and the repetition that used to go into mastering formulae is replaced by the safe repetition of harmless games. Proof never puts in an appearance, for fear that it will be discouragingly hard (and so waits menacingly in the wings, or is finessed altogether). But when children ask why, it ill serves them to answer: because it's nifty.

This benign approach has a second source, in political correctness—as a canny fourteen-year-old explained to us. He showed us a word-problem from his final exam: "A charity dinner was held for 1,500 students and adults. The adults were charged $15 per ticket, the students $12, and $21,000 worth of tickets were sold. How many adults attended, and how many students?" We said that this looked like a problem that might be solvable. "It's solvable all right," he answered, "but that's not the point. Look at how it's put: 'a charity dinner'. You're not allowed any more to talk about people making a profit." Of course! How naïve of us! The young must be screened away not only from difficulty and perplexity but from the fallen world of adults. It isn't that abstraction is replaced by reality, but that both are replaced by an "under fourteen"-rated video game. This systematically condescending attitude toward people in general and the young in particular has taken its toll in math teaching. By stifling how we imagine others and sugarcoating their differences, we injure imagination itself; by feigning compassion while enjoying the moral superiority it makes us feel, we nurture hypocrisy, which lames our relation to truth; by keeping students from ever feeling frustrated, we prevent them from having revelations. When, however, in The Math Circle, we accept any conjecture our students offer, this doesn't mean we endorse it, nor are we saying that your truth is all right and so is mine. Rather, the taking now obliges us to discover together whether this insight solves the problem—and the game's afoot.

## Ancestral Voices Calling for Reform

As if all these internal conflicts weren't enough, a medley of outside interests, at odds with one another, further distort the teaching of mathematics. These include pressures downward from working mathematicians; sideways from science, business, and industry; and in all directions from social concerns.

Although mathematics is deployed over a great expanse, its practitioners know its anatomy well enough to trace the connections among its major systems. Neophytes, they argue, should learn the progressively richer communities of number first: the naturals, integers, rationals, reals, and complex numbers, along with the operations on them. This leads the student through arithmetic to algebra, with geometry studied next or simultaneously (depending on the emphases in the mind of the person you're talking to). Now calculus, the fourth great system, opens up, and beyond it the spiral returns to arithmetic as number theory, and to the abstract algebra of groups, rings, and fields; and then (or concurrently) to the spectrum of geometries, and topology; and calculus rethought as real and complex analysis. The rudiments of logic, absorbed within some or all of the previous studies, resurfaces now in set theory, which is meant to slide under them all; and category theory arches overhead. This organic progression lays down the clearest possible guidelines for studying mathematics from the year dot.

The science lobby answers: this is a curriculum for pure mathematicians, and less than 1% of the students who start to climb will ever see the view from the top of that tower. What's needed instead is the ability to make sense of scientific data, and to think your way through real-world situations: knowing how to model a problem; understanding statistics and probability; being at home in calculus and, at its upper limits, with differential equations.

"And what percentage of students, these days, will be scientists?" ask the representatives of business and industry. This elitist curriculum ill prepares people for the workplace, where what you need is common sense sharpened by the discipline of mathematics (following rules to results, and knowing how to test them; doing word-problems). You need the sense of order in your life and thought that arithmetic gives—and all these innovations model only chaos. Teach them to add and subtract! Get students to learn how to use their calculators! If they're going to get ahead in the world, teach them programming!

Meanwhile, teachers have to think about accreditation from local, state, and national agencies; about climbing career ladders and winning merit pay. Their students are exhorted not to be left behind— rather to be accelerated, while not standing out from their group. Haven't we here the problem of Rousseau's *Emile*: the education of the free individual is contrary to the education of the useful citizen? Should math therefore be taught as a liberal art, or as a practical skill? Should the pain that its students and teachers suffer be respected, and the teaching of it accordingly cut down, or back, or out altogether (now that computers can do it all for us)?

"I wasn't doing anything and she said come to the blackboard and do this sum."—An indignant seven-year-old

And the mathematicians answer, as Plato did, that one studies mathematics to turn the soul's eye from the material world to the subjects of pure thought, and an a priori knowledge of eternal objects.

And the scientists reply . . .

These accumulating pressures periodically erupt in a survey that shows just how low the standards have fallen, and then (writes George Walden), "everyone slips into a well rehearsed posture . . . the unions call for cash; the teachers sulk; the press trumpets its outrage; another segment of the middle classes grits its teeth and stumps up for private education." And a committee meets to frame yet one more curriculum reform. Who is on this committee? In one case, in 1987, it was "nine mathematics educators, three school heads, four educational administrators, two academics, one industrialist, and one member of the New Right."

Most reforms address how math is taught, but a notable exception would change the math itself. Let's call this movement Muscular Mathematics; Back to Basics is a typical example. Enough of investigation, experimentation, discovery, and invention, say its proponents; enough of theorems and proofs, and "maybe it's the other way": replace ambiguity by certitude! Anything can be learned and made rock-solid by drill, and reinforced by Mad Minute quizzes, in which the aim is to get a hundred calculations perfect. Algorithms decode the universe, and minds are made to memorize them.

If you ask teachers what reforms they favor, they speak again and again about altering the way they themselves were taught. Students at a teachers' college were asked how they felt about math: "totally devastated"; "really useless and frustrated"; "stupid"; "humiliated"; "ashamed"; "inadequate"; "small"; "like crying"; "slow and thick"; "a failure"; "an idiot"; "shattered"; "depressed"; "bored".

Graduates of the college, who were now student teachers, were asked the same question: "It was all too fast for me—I couldn't keep up." "I just learned the rules in order to pass the examination." "By the time I got to the final year of my schooling, the gaps in my knowledge were so wide I gave up." "I don't have the basic mathematics knowledge to risk giving children very challenging work." "The word math makes me have a panic attack."

And what did veteran elementary school teachers answer? "Embarrassed"; "struggling"; "failing"; "terrified"; "demoralized"; "pressured"; "frightened"; "out of my depth".

When happy and successful teachers, who feel comfortable with the material, are asked for their secrets, they tend to talk geographically. Some prefer being in front of the room, where all their students can focus on them; some like leading from the rear, giving the students a sense that they are actually in charge; some like wandering about, helping this one here, consulting with another there, now leading, now following. What has location to do with it? Clearly nothing, since all positions work equally well; but like the knight who was given a dragon-taming formula, the teachers succeed in part because they are confident that they will.

What do teachers know anyway? The spirit of reform takes their ignorance as its premise ("In England," says a 1991 report, "the public acceptance of a national curriculum followed from a growing belief that teachers were both incompetent and politically suspect." The inspiration for and implementation of No Child Left Behind is based on the view that education is too important to be left to teachers.) School boards realize that their schools are failing, and since it is obviously impossible to replace all the math teachers, they buy a curriculum designed to retrain and support them. But just as every Torah becomes a Talmud, so every curriculum writer begins with a list of crucial topics, adds charming problems to tease attention and enthusiasm awake, and ends by writing out step-by-step instructions that explain nothing, yet leave nothing to the teacher's initiative or imagination.

Because the school board is answerable to the voters, there must be scientific means to test a curriculum's validity; these ways are, of course, statistical. What works for the majority must be used for all—hence no

HA!

134

more individualized instruction. Yet what actually happens in a class-room? There are some students whose intuition is geometric, others al-gebraic. There are eye learners, ear learners, and hand learners. There are those who need to work it all out for themselves in silence, and those who need company and conversation to stay focused. These are inde-pendent variables, so you'll find all sorts of combinations sitting in front of you. The teacher's job is to get everyone to master the ideas by finding the right combination for each of these complicated locks—but if she has to "follow the curriculum" rigidly, her hands are tied.

*Send to Juanita?*

The situation is, to be sure, much more complicated than this. What tone should you take to correct mistakes? Two researchers in 1976 found that in upper-income classrooms, praise was "negatively related to stu-dent learning gains"; in low, students prospered from a warm, support-ive approach. There's the culturally enhanced distinction between male and female styles of learning and taking part; there's the tenth-grade ambiguity splitter: some hate it, and opt for clarity and authority; some suddenly wake to the attraction of metaphor, and to the appeal of nu-anced answers. Each teacher of any experience will have a catalog much longer than this of the ways students differ in their outlooks and needs.

The inevitable solution of reformers—explicit for at least the last forty years—is to make curricula teacher-proof. Threads leading to this con-clusion in fact trail back at least as far as the 1920s, with those answer books, and manuals for teachers, and handy guides: meant to support, they end up crippling by keeping a person from exercising his mental muscles. Why not then make a virtue of this necessity and relieve the teacher altogether from having to stand up? Is anything more pathetic than a teacher, supposed to be guiding children's self-motivated investi-gations, who comes to rely on problems gleaned from old teachers' maga-zines, and hands out dog-eared cards with outdated information and says "Here's something new to think about"?

The next step is to suit textbooks carefully to the ability level of the classes they are to be used in, so that the teacher needn't have that respon-sibility. The School Math Project, for example, published "Y" texts for the ablest, "G" for the slowest. They differed not just in the difficulty of the math but in the attitude taken toward it. To paraphrase from a 1991 Open University Course: the "G" materials were *centrifugal*: lightweight in form and written in explicit, everyday language. They were mildly sensa-tional, and pictures outweighed text. The jokes in them were nonmathe-matical, and even antiacademic. The "Y" were *centripetal*: the language was esoteric, the jokes cerebral. The cover pictures celebrated the enig-matic and erudite (contour maps of the human face, Escher prints). "The mundane inevitably enters the domains of both series," this course con-cludes, "but whereas, in the Y books, the everyday world is sacrificed on the altar of mathematics, in the G series the everyday appears as an effusive apology for the tentative intrusion of the academic."

Such efforts take a considerable burden off the shoulders of a struggling teacher, but opportunities to interfere still remain. Why should the towered sorcerers (who, for all their powers, can't be present in all classrooms at once), let their apprentices wreak havoc with the wands put in their clumsy hands? This thought is taking us straight toward the pet goat. You may recall that President Bush was reading the story of a girl and her goat when he was stunned by the news of 9/11. It was written for second graders by Siegfried Engelmann, as part of his Direct Instruction method, which (according to a *New Yorker* "Talk of the Town" of July 26, 2004), dictates "every word of every lesson, including which words of encouragement teachers may and may not use." This article quotes from the Direct Instruction website: "The popular valuing of teacher creativity and autonomy as high priorities must give way to a willingness to follow certain carefully prescribed instructional practices." And it quotes Engelmann himself: "We don't give a damn what the teacher thinks, what the teacher feels. On the teachers' own time they can hate it. We don't care, as long as they do it." Direct Instruction's principles have naturally been applied to mathematics too. According to Engelmann, says the article, his is one of the few methods that has been consistently shown to improve student achievement. This brings it in line with the government's program, which requires that only scientifically based educational programs be eligible for federal funding. Perhaps there has been a paradigm shift in the basis of science.

Even when designing less extreme teacher-proof curricula, an inevitable consequence is that the texts become learner-proof too; the problem-writers so want to guarantee that their unseen students will succeed, that they can't leave them to figure out relations for themselves, but merely check them. Problems intended to foster discovery are given in such small spoonfuls that you needn't see the idea behind them at all (or even realize there was an idea) in order to answer each question in the sequence. Only the most aggressively obtuse student will fail to see that the sequence of hints converges to the answer.

Since the teacher has been written out of these scenarios—made superfluous even before being made redundant—the teaching machine follows as night follows day. Your patient and nonjudgmental friend is quietly blinking in the corner, awaiting your command. Here is one company's description of its paragon:

> The Company is currently completing the development of its first game, designed to teach Algebra I. The product seamlessly integrates math equations and concepts into an exciting first player

action adventure scenario that engages, encourages, and ultimately pulls students past the pain of learning. The game awards points for efficient problem solving, demonstrated improvement, and collaboration with other students (when played in multiplayer mode). It also deducts points for simply guessing. The parent sponsored points are integrated into the proprietary rewards system that allows students to exchange their points for merchandise and other rewards from strategic retail partners.

With this beside you in the wilderness, what more—other than perhaps a keg of beer—could you ask for?

## Anticurriculum

From this account you may think that curricula, like water, run only downhill—but there are movements against this flow, among which we count The Math Circle's predecessors. Perhaps their common ancestor is Rousseau, for whom the task was to develop what already lay in the glory-trailing child (and so gave rise to the false etymology of "education" as a "leading out"). Mathematics is free: if only the individual were, we would be living in an earthly paradise.

Johann Heinrich Pestalozzi (1746–1827) and, over a century later, Maria Montessori (1870–1952) developed Rousseau's image of the child along different lines. Pestalozzi sought to join in education what Rousseau had broken apart: the free individual and the citizen in society. He spoke of education "as a form of action which allows each person to recognize his own individuality and make a creative work of himself." He was repelled by systems, and focused not on theories of education but on the particular child's immediate concerns, which he addressed along the three routes of heart, head, and hand. The heart gives us our sense of union with our fellow men; the head, detached reflection; and having felt and thought, our hands now let us act to create ourselves—but all three must be equally involved in any educational undertaking.

Montessori saw the importance of focusing the child's attention and having hands shape what the mind sought for, the abstract becoming vivid in tangible geometric shapes. Her work with orphans made her realize that in an increasingly fluid society, schools must take on what was once the province of parents: bringing up children to comprehend others and to be comprehensible themselves. Her classrooms, like Pestalozzi's, fostered a collegial spirit, as do ours.

The need for deeper and more satisfying approaches to mathematics than could be found in schools led, in the last century, to all sorts of math clubs, math camps, math programs, and math contests, all with wildly differing agendas—some embroidering on, some part of, the fabric of mathematics. The University of Utah math circle meets during term time, and makes a particular effort to integrate both local high school students and their teachers with the faculty and students of the university's math department. There are distance learning programs, where problems are set and corrected via computer. Serious summer programs, such as the late Arnold Ross's at Ohio State, David Kelly's at Hampshire College, Glenn Stevens's PROMYS at Boston University, and the peripatetic MathCamp, immerse their generally adolescent students in several weeks of intense classes, lectures, problem-sessions, discussions, and camaraderie. Like music camps, these draw and wonderfully develop students who are already thoroughly devoted to the art. The relationship between the students and their instructors, who are often only a few years older, adds humanity and life to what might even be a fairly standard classroom format: exposition, problem sets, and presentation of results.

One very different approach to training young mathematicians was what came to be known as "Texas Topology," taught by R. L. Moore from 1920 to 1969, at the University of Texas.

**The Curate's Egg**

"I'm afraid," said the Bishop, in an ancient *Punch* cartoon, "that your egg is bad." "Oh no," answered the deferential curate, "parts of it are excellent." The Moore Method was R. L. Moore's egg. It too was excellent, in parts.

Moore began by interviewing each applicant. Any student who had already taken a course similar to the one he was applying for was ruled out; the aim was to have the whole cohort start from scratch, ignorant of established terminology, notation, methods, and results. In The Math Circle we always choose a topic that none present are familiar with (easier to do, certainly, with younger students), but we take whoever applies. Another difference from The Math Circle is that Moore also ruled out black students, and, they say, foreigners and Jews. On at least one occasion he refused to begin a class until a black woman in the room left.

Once in the course, the students were put on their honor not to read anything about the topic being studied. It says something about Moore's own sense of how honorable they might be that he removed all pertinent books from the University library. We, too, hope that students won't bring in ideas they've found online or learned from their parents. It adds an extra fillip to work around such difficulties on the spot: keeping up a collegial tone while getting across that the pleasures of discovery outweigh those of borrowed glory.

At Moore's first class meeting, the eminent mathematician, Paul Halmos, writes: he "would define the basic terms and either challenge the class to discover the relations among them, or, depending on the subject, the level, and the students, explicitly state a theorem, or two, or three. Class dismissed. Next meeting: 'Mr. Smith, please prove Theorem 1. Oh, you can't? Very well, Mr. Jones, can you? . . . ' If no one could, class dismissed." Well, not exactly. Moore wouldn't talk about the theorem thus left in suspended animation, but around it: about logic, or kinds of proof, or even history or just plain gossip—but hints toward the proof might be hidden in what he said.

Moore's whole effort was directed at the often nerve-wracking business of *proving*, which is why he would hand out, at the beginning of a course, the sequence of theorems that were to be tackled (and by proving them, master the subject these theorems marked out). Here, too, our approach differs, letting students work as mathematicians do, by first shaping a conjecture from inchoate data.

Presently, says Halmos, Moore's students were proving theorems and watching the proofs of their fellow students like eagles. You weren't allowed to let a mistake get past; it was your duty to point out mistakes to the presenter and demand a correction or offer one. Students, he says, were quickly ranked by quality, and once the order was clear, Moore would begin by calling on the weakest student. This prevented the best from holding the floor, and made for fierce competition, which Moore encouraged. "Do not read," says Halmos, describing the ethos of Texas Topology, "do not collaborate—think, work by yourself, beat the other guy. Often a student who hadn't yet found a proof of Theorem 11 would leave the room while someone else was presenting a proof of it—each student wanted to be able to give Moore his private solution, found without any help."

Needless to say, the bitterness of this competition goes dead against our grain. What's wrong with a vision of the little society of a classroom (or any society, for that matter) as a trading place for insights?

Unquestionably, Moore and his method turned out a vast number of PhDs in mathematics, many of whom became noted researchers and teachers. "Moore never missed a chance to praise students for their accomplishments," writes Peter Renz, "and worked tirelessly to bring out the best in all his students." One of Moore's biographers, Steve Kennedy, says that "his real genius lay in his ability to inspire people to do more than they themselves thought they could," and remarks that despite the competitive air, Moore's students admired, respected, and even loved one another. "Partly of course," he adds, "this is the foxhole phenomenon—we always remain attached to folks with whom we've shared an arduous, stressful trial."

Halmos again: "Moore felt the excitement of mathematical discovery and he understood the relation between that and the precision of mathematical expression. He could communicate his feeling and his understanding to his students, but he seemed not to know or care about the beauty, the architecture, and the elegance of mathematics and of mathematical writing." Moore was intolerant, he says, of every part of mathematics other than his own, seeing algebra and analysis as competitors and enemies; and made his students much less well educated and useful than they could have been.

And yet . . . and yet . . . Moore shared the same admiration that we, and all sensible people, do for the old Chinese proverb: "I hear, and I forget; I see, and I remember; I do, and I understand."

The large, dysfunctional patriarchy—what a very American story! A story, too, that calls Ty Cobb to mind: perhaps the greatest hitter of all time, who let it be known that he filed his cleats, so that the fielders would stay out of his way as he slid, spikes high, into the bag.

**Russian Math Circles**

We chose "The Math Circle" as the name for our approach to honor what we understood to have been a loose tradition in the Soviet Union of informal, often covert, meetings of students with teachers (their name, "circle", also made us think of those Decembrist philosophical circles, where conversation and debate flowed in amicable surroundings: free minds at play in an otherwise dangerous world). The image we had was of evenings of fervent discussion in someone's apartment, one person stirring the soup while another watched by the window for signs of the KGB. This picture may not have been wholly accurate. In fact these circles may have had a Hungarian origin, in 1894, with competitions that came to be called The Hungarian Nursery—and a spirit of competition has informed them since. The contestants work in classrooms under supervision, the Society selects the two best papers, and the awards—a first and second Eôtvôs Prize—are given to the winners by the president himself at the next session of the Society. Notice, however, that even in Hungary some have had doubts about the value of competition. In Gábor Szegú's 1961 preface to the *Hungarian Problem Book,* he said that although competition was a powerful stimulant, it was not necessary for a lively mathematical culture, and that the students who were so deeply involved in the long and hard work of solving the problems were probably spurred on not by the possibility of a medal, but by "the attitude which rates intellectual effort and spiritual achievement higher than material advantage"—something that couldn't be fostered by decrees from on high or more and more intense mathematical training: "the most effective means may consist of transmitting to the young mind the beauty of intellectual work and the feeling of satisfaction following a great and successful mental effort."

In the Soviet Union, at any rate, math circles went back at least as far as the great mathematician Andrei Nikolaevich Kolmogorov (1903–87).

This looks like Carl Sagan.

In addition to his own vast research (it was said that it would be shorter to simply list the fields in which he *hadn't* done significant work), he was deeply committed to teaching young students. When he taught math he didn't lecture but just posed problems, gave the students complete freedom to attack them as they saw fit, and as his biographer said, was "always waiting to hear from the student something remarkable."

Despite the intensity of his own passion for mathematics, he encouraged his young students to venture into other realms, introducing them to "literature and music, joining in their recreations and taking them on hikes, excursions, and expeditions." His idea was that students shouldn't be intellectual clones of their teacher but should develop their own personalities in life and in their work. He didn't even mind if they didn't become mathematicians, as long as they retained the principles he taught: a broad outlook and an unstifled curiosity.

Russian math circles and math competitions meshed imperfectly, like large and slightly worn wooden gears. In Moscow, at least, both were organized by the same people, though the students involved with one needn't have been (but usually were) involved in the other. In 1934 Moscow State University (MSU) started a series of lectures for high-school students and a series of math publications aimed at them; in 1935 the first Moscow Math Olympiad was held, with 314 participants, and the next year a math circle was organized at MSU. Students were bored by professors' long lectures and students' clumsy presentations. In the late 1930s this changed radically: an MSU student named David Shklyarskiy organized his section as a short lecture, followed by students' discussing solutions to problems they had worked on. The problems were of varying degrees of difficulty and could have something to do with the lecture—or not. Some were mini-research investigations. Occasionally Shklyarskiy asked students to find elementary solutions to problems for which none were known, and finding these solutions sometimes extended over years. The extraordinary success of his students at the 1938 Math Olympiad, where they won half the prizes, led to Shklyarskiy's format being adopted nationwide and maintained into the 1960s. By then some math circles focused on problem solving, and others on discussing a weekly topic (these were called "mathematical etudes", with an attractive nod to music).

The body of mathematics may be universal, but it wears the clothes of a time and place. In these Russian math circles students always called the teacher by his first name rather than his patronymic: rare in Russia for someone addressing an older person, and the teacher always called students by their last names, as if they were already mathematicians. At the first meeting of a math circle, the various teachers would give short advertisements for their courses, promising everything except success in the Olympiad. "In our section," one would say, "we will do everything that previous teachers have already talked about, but also . . . " Section size varied from three to more than a hundred, and if a section had too many students in it, the presentation was simply made more difficult, encouraging students to leave. Why does this alternating attraction and repulsion remind one so of Dostoyevsky?

The Moscow Math Olympiads attracted thousands of students from all over the Soviet Union. Its two rounds extended over four spring Sundays: in round one the students spent four or five hours attacking four to six problems; the next Sunday there was a public discussion of solutions and common mistakes; then the successful candidates—the 30 to 50 percent who had solved at least two of the problems—went on to round two, where they were faced with another four to six problems, but of much greater difficulty. On the final Sunday, the winners were announced and presented with their prizes: not cups or plaques, but autographed copies of small math pamphlets.

In the spirit established by Kolmogorov, a student who hadn't solved a single problem might yet be a winner, if his work showed genuinely mathematical originality, thoughtfulness, and rigor. In 1945, for example, solving a relatively simple geometry problem depended on recognizing that a line which doesn't meet any of a triangle's vertices cannot intersect all three of its sides. The Olympiad committee thought of this as something that students could take for granted. But the fifteen-year-old Roland Dobrushin got to this point, then wrote: "I have spent a long time trying to prove that a straight line cannot meet all three sides of a triangle at internal points, but I couldn't do it, since I realize to my horror that I do not know what a straight line is!" For this confession, he was awarded first prize, since this is indeed left as a mystery in Euclid's geometry.

All those students from every part of the vast Soviet Union! Let's lift the veil of anonymity and look at one. In 1963 Tatiana Shubin was a thirteen-year-old in Alma Ata, in Siberia, when a math circle was set up at her school. There was already a dance circle, an art circle, a theater circle, and so on. The math circle lasted for only a few sessions, because all the students did was to read short biographies of prominent mathematicians. The next year Tatiana saw, in the *Komsomol'skaya Pravda*, an announcement about the first stage of the all-Siberian Math Olympiad with a collection of problems, arranged according to grade levels; readers were encouraged to send in their solutions. Tatiana tried them and was invited to the next level of the competition, which was to be held in her hometown. The participants were seated in different classrooms, according to grade level, and were given several problems to solve in three or four hours—with the inevitable proctors vigilantly patroling the aisles. The top scorers from this competition were invited to a summer school in Academgorodok, a town newly built in a pristine forest, across the Ob' river from the great city of Novosibirsk.

The summer school lasted a few weeks. Tatiana went to daily lectures—some given to the entire camp by leading research scientists, some more specialized and held in smaller auditoriums by younger and less famous people. "I was really taken," she said, "by a series of two or three lectures in which relativity theory was explained—mostly by means of diagrams. The clarity of it was simply breathtaking." There were also problem-solving sessions led by graduate students, but no activities were compulsory, and many students preferred simply to roam the streets of this town of science and scientists, sheltered by the magnificent Siberian forest.

At the end of the summer there was a final competition, and those who succeeded were admitted to the Physics and Mathematics Boarding School, which, because it was a department of the Novosibirsk State University, escaped the scrutiny of the Ministry for the People's Education and the severe limitations it placed on education elsewhere. Tatiana started studying there in the golden autumn of 1965.

It was all very different from a standard school. There were big lectures, given by university professors, and small problem-solving sessions led by graduate students. In other schools, every student in the nation was taking the same set of courses and using the same textbooks. Here the state curriculum was ignored. In math courses everything was proved rigorously, and students were expected not only to remember the theorems and their proofs, but also to devise their own proofs in the problem-solving sessions. Their homework exercises were like Olympiad problems, and required a lot of hard thinking. For physics, the students took part in the real work going on in the labs, in a building with an impressive brass plate at the door, inscribed "The Siberian Branch of the Academy of Sciences".

Tatiana could have stayed there until she graduated, but after a year her parents wanted her to come back home. That time at the Physics and Mathematics Boarding School—its freedom, its intensity, its remoteness, farther than Academgorodok, in a world long since disappeared—shimmers still under the dark pines. A story without the expected happy ending.

# seven

# How Mathematicians Actually Work

We spoke at length in chapter 4 about what goes into mathematical work. Here we would like to step back just enough to consider how these ingredients blend together when we wrestle with those unsolved questions that no teacher's manual can help with, which are (in the quiet terminology of the trade) "open". To do this fully, we would have to provide, or presuppose on the reader's part, the knowledge needed to reach these sectors of the front line; otherwise you would just have a Monty Python sketch of famous mathematicians furrowing their brows and then shouting "Eureka!"

Our aim is much more realistic. Not "real world" as in balancing your checkbook—machines are much better at that than we are; even making change is fast becoming a lost art as coins themselves disappear. Instead, math teaching should focus on what humans do best: using reason to understand the underlying structure of things. That's what mathematicians do. We don't want our students to suffer through dispiriting rigors imposed by the changing fads of educational theories, but to approach mathematics, *from the beginning,* as working mathematicians. For in this way they will be apprentices in the guild, where making is all of a piece with learning, and the ins and outs of the craft are collegially shared. We need to have at least a first approximation to what this work is like: how the assorted tools we have described are laid out, used, modified, discarded, or added to. The peculiarities of personality are always important in math as in any other creative work; an open problem drifts through the profession, encountering different temperaments, emphases, and inclinations of thought, until it meets the mind that fits it. There is no single mathematical type to which students must be shaped, and this is exactly as it should be; we want to create those conditions that will let the greatest diversity of personalities prosper. Mathematics is freedom—within the constraints of a certain kind of thought. That "kind"

The real mathematician is an enthusiast *per se*. Without enthusiasm no mathematics.—Novalis

is the genus to which we hope the widest speciation of individuals will belong. It is ultimately the human kind—but with certain acquired habits and dispositions.

How is it, actually, that mathematicians work? And why do we think that all students should learn to work that way? There's a lot to be said for the received opinion of how mathematicians practice their art: data, relations among and speculations about them accumulate, and a deep point of difficulty slowly takes shape in the swirl. One approach after another fails, but each contributes its mite to the content and form of the mass. Then (you will recognize from the language of this explanation that it emerged in the first half of the last century) the work sinks below the horizon of consciousness but continues in the unconscious—or in its fringe—which is somehow more congenial to the making of significant combinations: perhaps because restrictions are relaxed that confine conscious thought too narrowly. Should the right combination take shape there, beyond your awareness, it will appear unpredictably, all at once, and you will wonder both how you saw it, and why you hadn't seen it before.

The *locus classicus* of this explanation is a 1913 lecture of Poincaré's. He begins by saying that he will be talking about his discovery of a proof for a theorem with "a barbarous name, unfamiliar to many, but that is unimportant; what is of interest for the psychologist is not the theory but its circumstances."

He then describes how he worked for two weeks on "Fuchsian functions", trying in vain to prove that there couldn't be any; an idea, he says, which was going to turn out to be false. He speaks of a sleepless night, during which he constructed one species of these functions and then tried to find an expression for them. "Ideas rose in crowds; I felt them collide until pairs interlocked, so to speak, making a stable combination."

> Just at this time, I left Caen, where I was living, to go on a geologic excursion under the auspices of the School of Mines. The incidents of the travel made me forget my mathematical work. Having reached Coutances, we entered an omnibus to go some place or other. At the moment when I put my foot on the step, the [solution to the problem] came to me, without anything in my former thoughts seeming to have paved the way for it.... I did not verify the idea; I should not have had time, as, upon taking my seat in the omnibus, I went on with a conversation already commenced, but I felt a perfect certainty. On my return to Caen, for conscience' sake, I verified the result at my leisure.

Lest you think that this remarkable process was unique to Poincaré, the mathematician Jacques Hadamard collected more examples from himself and others such as Gauss, who—writing about a theorem he had tried unsuccessfully to prove for years—says: "Finally, two days ago, I succeeded, not on account of my painful efforts, but by the grace of God. Like a sudden flash of lightning, the riddle happened to be solved. I myself cannot say what was the conducting thread which connected what I previously knew with what made my success possible." *Happened to be solved*, a wholly impersonal tone: neither Gauss nor God but "of itself".

It was a theory of the time that illuminations such as these are preceded by an "incubation" stage, during which effort ceases and the subject is dropped. Prior to this, of course, there had to be an initial stage of total immersion in the problem and its context: an identification so thorough that you couldn't imagine its architecture to be different from what it is. "If I understand mathematics," as Isaiah Berlin put it, "how can I want something which goes against it? To understand is not merely to accept, but actively to want what is understood." Or, as Hadamard said, the unconscious selects those ideas "which satisfy our sense of beauty and, consequently, are likely to be useful."

To speak of the agency of a sense of beauty, operating in the unconscious, raises perhaps more problems than it solves (how close it is to classical notions of the good and the beautiful being the same: καλοκἀγαθός for the Greeks, *decus*—fitting—for the Romans); but the "combination" metaphor has attracted much attention. It is wonderfully put by John Livingston Lowes, in describing how Samuel Taylor Coleridge came to write his magical poem, *Kubla Khan*:

> Suppose a subliminal reservoir thronged, as Coleridge's was thronged, with images which had flashed on the inner eye from the pages of innumerable books. Suppose these images to be fitted, as it were, with links which rendered possible indefinite combination. Suppose some powerful suggestion in the field of consciousness strikes down into this mass of images thus capable of all manner of conjunctions. And suppose that this time, when in response to the summons the sleeping images flock up, with their potential associations, from the deeps—suppose that this time all conscious imaginative control is for some reason in abeyance. What, if all this were so, would happen?

Those entangled atoms come from the pre-Socratic physical theories of Democritus and Leucippus; they might more accurately describe events not in the outer world of physics but the inner world of thought. We picture deductive chains catching on to one another, thens to ifs, or even attaching midway, as lemmas and corollaries will sprout on the way to a theorem, or those unexpected results that the Greeks called κέρδοι (shrewd gains) will drop like ripe fruit from a tree.

Evariste Galois, a century before this explanation took shape, objected to any idea of chance playing a part in mathematical discovery: *"La science progresse par une série de combinasions où le hazard ne joue pas le moindre role"* (science progresses by a series of combinations where chance plays not even the smallest part). Still, the theory of subliminal linkages satisfies the Cartesian desire for thought to move with élan, smoothly and swiftly. The sudden discontinuities caused by incubation's fallow periods, and the startling eruptions of discovery, could now be discounted, by attributing them to the continuous work hidden in the unconscious.

But in this attractive theory, there are more difficulties than finding the artistic hand in unconscious work: questions about "unconscious work" itself, since, by its very nature, it is inaccessible to objective scrutiny. How can we understand its character? And when an intermediate "fringe" con- sciousness is invoked, shading off to a sub-, then to an unconscious,* these difficulties are transported wholesale into it, along with the new one of *its* dubious status: fringed or not, what guarantee have we that this "waiting room" even exists? How does this explanation differ from those based on chance, inspiration, or divine intervention? Even if we were to argue that all possible, or at least all types of, chance combinations occurred in this deep well, with only the "right" ones somehow surfacing, might not so combinatoric a theory distort our view of the mind as much as putting combinatorics at the heart of mathematics would distort *its* nature, since other equally well-established enterprises thrive there?

Interestingly enough, a quality that this explanation can't help but stress (through the association of consciousness with volition) is that these miraculous solutions arise "all at once", seemingly without our will—remember Gauss's use of the passive voice. Such a quality, how- ever, may help us to understand the source of this explanation, and at the same time, undermine it. For might it not consist in illegitimately extending a sense of how we learn to how we discover?

The argument is this. Learning of any sort—from how to handle a new tool well to how to speak a new language fluently—moves from incessant use (in different ways and in different contexts) to a point when we retrospectively realize that we no longer have to practice but find ourselves used to it; it becomes a usage of ours, no longer an end but a means to other ends. What we've thus mastered now effortlessly pro- duces results. Rapid calculators, people who instantly know the time, or to which compass point they're facing, each of us when we bank our sentences flawlessly between grammatical slalom gates, knows that our will isn't involved—yet we resist conjuring up lucky entanglements of

---

*"Fringe" is due to William James. The notion of an antechamber to consciousness is, prior to him, in John Locke, and later, Francis Galton, and the Freudian "foreconscious", follows him.

nouns and verbs in our unconscious, or the agency of a just sense of time, space, or summation. The temptation to see mathematical invention that way comes from a problem's seeming to lie past the reach of whatever skills we have, whether conscious now, or formerly. So we revert to an animistic sort of thinking and picture the problem teaching us.

> The mathematician lives long and lives young; the wings of his soul do not early drop off.
> —J. J. Sylvester

Here is a different understanding of mathematical invention. As we cross and recross the landscape of our knowledge, with this sinkhole of a problem plunk in its middle, we become ever more familiar not only with the landscape but with the pathways themselves: shorter and longer lines of implication and association that branch and interconnect. They begin to separate into those we are conscious of having made (which therefore seem both artificial and private), and the older ones, which take on the air of having been found: of belonging to this more and more natural landscape. The paths harden into truths as compact as the "facts" they connect, so that a surface of impersonal content seems to solidify beneath our tentative probes and conjectures, which now grow into a new context for that content.

Problems are solved by relating them fruitfully to configurations we thoroughly understand, just as you can memorize anything by breaking it down into parts, and imagining each placed in the details of a house you know well—the Palace of Memory. This making of new relations can happen only when past relations thus simplify into objects for the new relations to connect. Nets in our understanding must contract to nodes, in order to lie in new nets. This is once again that recursive abstraction, so deeply stamped on mathematical thought, which we saw in learning mathematics by climbing its tall building, floor after floor—so that this explanation fits better than the first with the way mathematics as a whole evolves.

Instead of invoking an "unconscious", with its inaccessible turnings, this explanation calls up a layered consciousness, the old arena of thought made into a stable locale that the newer one surrounds with a relational, dynamic context—which in its turn will contract and be netted into higher-order relations. True, the central problem still lies open, since our work with those features that lay on the level of the problem couldn't solve it. This is why the likeliest solution comes from Mind so distancing itself that it can now work *on* rather than *with* these features.

The insight that unlocks a problem (this means seeing how its parts are fitted together or keying such an insight to a proof) may consist in drawing relations among a few or many of the mechanisms, or in wholly transforming much of the old content by rotating the diamond and seeing it from a fresh point of view.

We could re-invoke the notion of a "conductor" who oversees this rotating as well as the many other activities, on so many different levels, going on here: keeping track of this open question's slowly closing boundary, and of new questions bulging out from it; mulling over different styles of proof and connections to well-explored territory; judging how much of the visual to draw on, how much of the algebraic; what debts of proof are still owed to steps along the way whose truth was no more than provisional; and why, after all, this statement is not only so but *should* be so.

Let's risk here, instead, a guess at how this web of relations could be self-regulating. Human thought has been linear from the beginning: we are experts at making and following long causal chains. Are we now, perhaps, in the early stages of understanding non-linear thought: with how we could move, both locally and globally, in the wide spreading networks that are a more accurate representation of the mind's geography? Do the "inclinations", "intuitions", and "instincts" we casually speak of represent our evolving capacity to gauge which path or paths from a junction are better followed—and is this capacity no more than the weighted network itself, condensed to a mobile point of view within it—or even more simply, just those varying gravitational pulls?

And here we reach the central relevance of theory to the teaching of mathematics: coming to think in this layered way, not memorizing a lot of facts, is what makes a mathematician. How is it taught? Not by the dreary practices described in the previous chapter, which serve only to stiffen pliable thought into "do this when you see that"—but by involving mind and mystery in each other, their interplay unfolding each. Here are two representative miniature problems (which you can enlarge by worrying at them before looking at how they resolve).

A monk one morning starts to climb the path up a holy mountain and reaches the top by evening. The next day he takes the same path down. Is there any point on this path that he passed on the second day at *precisely* the same moment as when he ascended? You might ask: were his departure and arrival times identical? Not necessarily. He may have slept late the second day or stopped to admire the view on the first. Is the holiness of the mountain important, or his being a monk? Not at all: these were window dressing. Do we need to take sophisticated mathematics along as we follow this monk? Nothing more than sandwiches if you're hungry.

This is the sort of puzzle that leaves you convinced the world is unfair, or that you're just not up to its guile, or, if feeling self-confident, that a positive answer wasn't expected: there need be no such point, and were there one, we'd never be able to prove its existence or find it.

Yet if you change the context from *time* to *space*, the answer leaps out. Instead of a single monk on two days, make it two monks on the same day, one starting in the morning from the bottom, the other from the

150

*Good question?*

top. Of course they will pass each other at some moment—and thus be in the same place, at the same time.

The second example involves an outrageous answer to an outrageous question. By now you may be at ease with infinite sets, but it took immense strength of character as well as of imagination to change the authoritative picture of numbers a hundred and fifty years ago, when enumeration of course went on forever, but you couldn't speak about a *completed* infinite totality. Just as forming ratios preceded the idea of rational numbers, and subtracting made sense before negative numbers did, so infinite processes were legitimate—but not yet infinite wholes. In mathematics, at least, nouns come after verbs. The paradise of infinite structures opened up by this changed point of view is testimony enough to there being fully conscious and noncombinatorial ways of mathematical invention. It is in this context that our example lives.

Intuition tells us that even if we are comfortable with thinking about infinite sets, the infinity of positive rationals (let's call it $Q^+$) is greater than the infinity of counting numbers ($\mathbb{N}$), since each counting number is followed by a distinct next, but the rationals are dense: between any two there is always another, so that you go down rabbit holes to infinity everywhere along the number line. Yet suppose you suspect that in fact there are no more positive rationals than naturals—that a 1-to-1 correspondence could be made between $Q^+$ and $\mathbb{N}$: how could you confirm your suspicion?

We must certainly alter radically the rational landscape as we know it, since to make up such a pairing we would need to match the counting number 1 with a first rational—but there is no smallest (were you to claim that some a/b was smallest, a/2b would be smaller still). We must painfully peel apart the notions of "smallest" and "first". If we are to count the members of $Q^+$, we won't be able to do it by size. We need a different notion of "first".

Here's one that violates two more self-evident truths. You may have hoped, when young, that a/b + c/d = (a+ c)/(b + d), but soon learned that this was heresy (1/2 + 1/2 ≠ 2/4). The reality is much more awkward: a/b + c/d = (ad + bc)/bd. Well, now we're going to combine fractions in the heretical way: a/b and c/d are to yield (a + c)/(b + d). It must be that "and" is no longer "+" and "yield" is no longer "="; since we've just seen that this isn't how you *add* fractions. Rather, this way of making two fractions produce another is just invented to help us match up the elements of $Q^+$ with those of $\mathbb{N}$.

Since we're in an iconoclastic mood, let's shatter one of the oldest idols of all: you can't divide by 0, so that 1/0, for instance, makes no sense. Agreed. Nevertheless, mathematics is freedom, and our subversive conviction that, despite appearances, $Q^+$ and $\mathbb{N}$ are the same size, frees the power of invention in us. Let's make use of the *mere symbol* 1/0, along with our new way of producing fractions, to construct an ordered list of all the fractions in $\mathbb{Q}^+$, as follows.

Set up 0/1 and 1/0 as goalposts and let play between them begin according to the rule: a/b and c/d yield an intermediate (a + c)/(b + d). So 0/1 and 1/0 thus yield 1/1:

$$0/1 \qquad\qquad 1/0$$
$$1/1$$

1/1 will be the first rational on our list (the goalposts were there simply to start play).

Apply our rule again to find intermediates, first to the left, then to the right, of 1/1:

$$0/1 \qquad\qquad\qquad\qquad 1/0$$
$$1/1$$
$$1/2 \qquad 2/1$$

Our list is now:
1. 1/1
2. 1/2
3. 2/1

Continue, once again moving from left to right:

$$0/1 \qquad\qquad\qquad\qquad\qquad\qquad 1/0$$
$$1/1$$
$$1/2 \qquad 2/1$$
$$1/3 \quad 2/3 \quad 3/2 \quad 3/1$$

This swells our list to
1. 1/1
2. 1/2
3. 2/1
4. 1/3
5. 2/3
6. 3/2
7. 3/1

By going on in this orderly way, inserting new terms between the old, from left to right, each positive rational will appear on our list once, and only once—so that our list shows the perfect match-up of $\mathbb{N}$ and $\mathbb{Q}^+$.

When you think back over these movements of thought, you catch a strong whiff of what mathematical work can be like. It isn't that new configurations always flash suddenly across a benighted sky. J. J. Sylvester gives the example of Sturm's theorem about the roots of equations: "as he in-

*(from succession to coordination*

formed me with his own lips, [it] stared him in the face in the midst of some mechanical investigations connected with the motion of compound pendulums." An inclination, a guess, an intuition may loosen familiar structures and let you tinker with them until they smoothly reconform. And as always in mathematics, so thorough is this reconfiguration that it would be unthinkable now to do without the very thing that seemed unthinkable before—although it usually takes a while for the new order to be grasped as firmly as the one that it replaced came to be, during *its* tenure (language has to swerve around and catch up with thought).

Both explanations have us begin immersed in the problem and its setting, feeling at home everywhere but in the problem's neighborhood, where our familiarity seems not only useless but even possibly mistaken, if it can have a crevasse such as this in its surface. A tectonic landscape, then, revealing its severe strains. And in both explanations we distance ourselves from this landscape—now, however, not to the unconscious depths beneath it but sufficiently far above to see how to connect the connections down there, or even to reconfigure the whole landscape. Both actions move away from a linearly ordered path of thought to a branched ordering: from succession to coordination. This in turn may bring together regions that had otherwise been ineffectively far apart. So a painter, stepping back from brushing in a detail, will adjust his composition locally, and then at large, to balance its main masses.

The work that happens on this higher level might of course be combinatorial—but it can also flow smoothly between rocks and whirlpools of problems. It may end up creating a stereoptical kind of seeing that comes from developing rival hypotheses at the same time. ("Nothing is more dangerous than an idea," said the nineteenth-century geologist Thomas Chamberlain, "when it is the only one we have.") We might even seek both to confirm and refute an hypothesis; remember that the functions Poincaré created were those he was trying to prove couldn't exist.

We may see, from up here, that constraints we had thought held us in check don't in fact exist (in the nine-dot problem,

who said you had to stay within the square of nine dots, when trying to connect them with a four-segment line?).

We may confront algebraic and geometric points of view, or find a way to translate a problem in one part of mathematics into the language of another, where it parses more comfortably; or we may use probabilistic arguments to establish certainties. We may call to our aid mere fictions, which in the using take on surprisingly the gravitas of reality, and like the little Robin Goodfellow of the tales, grow from innocuous helper to the keeper of an unguessed world's secrets (we used 1/0 to order the rationals, just as on page 59 we showed Euler using complex numbers to unlock a riddle about real numbers; indeed, most of our deepest devices, from 0 to x, entered as no more than notational conveniences). The repertoire of constructive thought is at least as rich as that of musical composition, where modulations may happen by jumps, slides, or bridge-passages, or so gradually as to be all but unheard.

Small wonder that this creative thinking is difficult to describe, since language moves from point to point within its own linear grammar. Here is another description to compare to ours. William Thurston, whom you heard from before (p. 100), suggests that mathematicians don't think via long deductive chains, but break proofs into more easily analyzed bits—and then

> have amazing facilities for sensing something without knowing where it comes from (intuition); of sensing that some phenomenon or situation or object is like something else (association); and for building and testing connections and comparisons, holding two things in mind at the same time (metaphor).... Personally, I put a lot of effort into "listening" to my intuitions and associations, and building them into metaphors and connections. This involves a simultaneous quieting and focusing of my mind. Words, logic and detailed pictures rattling around can inhibit intuitions and associations. We tend to think more effectively with spatial imagery on a larger scale: it's as if our brains take larger things more seriously and can devote more resources to them.

The problem deserves its share of the credit. Just as matter deforms spacetime in its vicinity, and functions are globally defined by where their singularities are, so problems deform and define the curvature of the mathematical context we have built up in our minds. This may seem paradoxical, since mathematics aims to present a coherent and universal picture of the world which mind inhabits. But that picture grows from a hundred troubling and therefore germinal locales, those particulars we spoke of on each of which reads as an exception—trying both the rules and our patience.*

---

*The history of mathematics, and one's own mathematical biography, are woven around these exceptions. Ancient lists would have had the five regular polyhedra on them, for example; the current list includes the sporadic simple groups and the exceptional Lie groups. John Stillwell says of them: "In the mind of every mathematician, there is tension between the general rule and exceptional cases. Our conscience tells us we should strive for general theorems, yet we are fascinated and seduced by beautiful exceptions." Any problem is an exception to the current state of our knowledge. Think of the problem about the number of regions into which n chords divide a circle. We understood the sequence 0, 1, 2, 4, 8, 16 perfectly well, until the next number, 31, opened the door to a wholly different sort of counting.

Barry Mazur, whom we've quoted before, conceives of the creative process in a way that combines provocative objects with conscious discovery:

> I think of it as a soup you keep stirring. You don't understand why it never comes to a boil. Every once in a while a bean jumps up in it and you dismiss it as uninteresting. Then it pops up again—and again. Finally you say to yourself: "I've got to hang on to that." The key to understanding is a shifting evaluation: recognizing that the pathological annoyance isn't peripheral but central. You'll have to work through this puzzle before you'll be able to get any further—so that the evaluation of your priorities also shifts. If you go on thinking of this problem as pathological, it remains just that, but once you say, well, I'm not understanding anything fully, in this soup—then things are going by too fast. Perhaps I'd really better understand this problem, this calculation I've done a hundred times: understand it better than I ever have.
>
> Isn't that really what happened to Poincaré? It's like moving your head to see the same figure from another viewpoint, and realizing that you have a much greater range of mobility in your neck than you'd thought. It is all there, and has been there for a very long time, but all of a sudden you appreciate an undervalued aspect of it as important. *Edward de Bono & his "PO".*

This view suits with an observation from the theory of evolution, that the new dominant species will often arise not from the former second best but from an altogether neglected and backward strain, which now thrives in an altered environment whose alterations enhance and are enhanced by the new species' unexpected growth. It also highlights the play that Stillwell spoke of: at the same time that abstraction carries thought away to the ever more general, the contrary motion of imagination arrows down toward rich particulars, and these extremes cross-pollinate.

> The moving power of mathematical invention is not reasoning but imagination.—Augustus De Morgan

We left it to a commentator on Coleridge to embody the received view that we began with. It seems only right to let A. E. Housman describe, in terms of poetry, where we've ended:

> I think that the production of poetry, in its first stage, is a passive and involuntary process; to name the class of things to which I think it belongs, I should call it a secretion; whether a natural secretion, like turpentine in the fir, or a morbid secretion, like the pearl in the oyster. I think that in my own case it is the latter; because I have seldom written poetry unless I was rather out of health, and the experience, though pleasurable, was generally agitating and exhausting.
>
> Having drunk a pint of beer at luncheon—beer is a sedative to the brain, and my afternoons are the least intellectual portion of my life—I would go out for a walk of two or three hours. As I went along, thinking of nothing in particular, only looking at things

around me and following the progress of the seasons, there would flow into my mind, with sudden and unaccountable emotion, sometimes a line or two of verse, sometimes a whole stanza at once, accompanied, not preceded, by a vague notion of the poem which they were destined to form part of. Then there would usually be a lull of an hour or so, then perhaps the spring would bubble up again. I say bubble up, because, so far as I could make out, the source of the suggestions thus proffered to the brain was the pit of the stomach.

Keep in mind that the oyster's secretion is irritated into activity by a speck of impurity—which plays the role in Housman's telling that an exception does in ours.

*　*　*

As we turn, in the next chapter, from talking about how mathematics is made by mathematicians, to how mathematics makes mathematicians, something of what we've seen along the way should keep us company.

**Time.**

What Poincaré's story certainly illustrates is the emotional spaciousness and leisurely sense of time that mathematical thinking thrives in. Working within converging walls of time may funnel intensity down, but makes what it works on seem an arbitrary goal rather than an entrée to significant vistas. The chance to profit from being wrong has been squeezed away, and we're left with the illusion that solutions are always sailed at head on, rather than tacked around. How easy it is to forget that mathematics is a humanistic rather than mechanistic pursuit.

*Nice point*

**Company.**

Moving the seat of creativity from the unconscious removes as well any desire to invoke a "collective unconscious". Collective consciousness, instead, seems an appealing backdrop against which the idiosyncrasies of personality can play: a generous sharing of insights in collegial conversation. Now the tacking of minds around a question weaves what may be a sufficiently dense context for catching its answer. These outer exchanges evolve with the inner voices of each. And as the unconscious loses its explanatory power, so does William James's notion of its fringe. Shall we replace it with his brother Henry's image of the web he would have every writer spin, responsive to the least passing nuance?

**Messing about.**

Thurston's remark about the variety of means at our disposal reminds us again that only machines have methods: we have approaches, with all the ad hockery they entail. We improvise, we make do, we work *from* principles, *by* analogy, *toward* conclusions, *with* algorithms, we devise a tool for just this little adjustment, leap from one association to another and creep down deductive paths, we remember and forget—but we lean at our peril on memorizing, since that only ossifies the surface of sense.

**The pendulum between the general and the particular.**

The spirits of Hilbert and Ramanujan lean over our efforts: the one ever lifting us up toward the form of the whole, the other dipping down again and again to catch at the invigorating singular.

> Beautiful general concepts do not drop out of the sky [Hermann Weyl once wrote]. To begin with, there are definite concrete problems, with all their undivided complexity, and these must be conquered by individuals relying on brute force. Only then come the axiomatizers and conclude that instead of straining to break in the door and bloodying one's hands one should have first constructed a magic key of such and such a shape and then the door would have opened quietly, as if by itself. But they can construct the key only because the successful breakthrough enables them to study the lock front and back, from the outside and from the inside. Before we can generalize, formalize and axiomatize there must be mathematical substance.

This stirred soup is a spiral nebula, exceptional in each of its stars.

**Simplicity.**

"The solution stared him in the face"—and when you see it, you see it simply, as a whole: a face staring back at you. It may be this simplicity coalescing out of chaos that Hadamard read as the sense of beauty, choosing the right one among all those chance combinations. In working on a deep problem, simplicity is the ultimate guide, letting us understand that we understand.

# eight

# The Math Circle

> In most sciences one generation tears down what another
> has built and what one has established another undoes. In
> Mathematics alone each generation builds a new storey to
> the old structure.—Hermann Hankel

It is as true of The Math Circle as of any piece of mathematics that what happens in it follows from its axioms. Ours is: mathematical content grows best in a congenial context.

Mathematical content: not finger exercises or busywork but the real stuff of mathematics—significant structure discovered in the world and proofs for it invented in the mind. "Structure" includes what *can't,* as well as what *must,* be the case.

Congenial context: the serious play of a few minds together, enjoying the common pursuit and each other's efforts, without the external compulsion of time or the internal compulsion to triumph, so that the architectural instinct may flourish in each (for this instinct is the middle term between world and mind).

Let's flesh this out.

We aim to waken in everyone first an awareness of, then a love for, and finally the power to do mathematics. The means seem straightforward: putting an attractive nugget of math in front of people and letting them play constructively with it. But as a reflective observer once said of a Math Circle session: "How did you do what I just saw?" For a lot of decisions have to be made beforehand about every noun, adjective, adverb, and verb in this paragraph's first two sentences. Why should teaching math be harder than teaching anything else? Why should it be harder than teaching someone to ride a bike? Because there is no skeleton key to unlock different topics and different personalities. Mathematics is layered—and so are people. They have to be riffled thoughtfully together.

With the previous chapters in mind, we'll talk now of our ends and beginnings, the math and the students—keeping in mind as well how interleaved they are.

## Ends

Once the ends are clear, the means click into place, like synapses. But these ends need to be imagined in sufficient detail so that, like reasoning backward, the path to them will take shape. Let's be more precise, then, about waking up awareness, love, and power, by thinking again of the mathematical enterprise as a landscape.

In its foreground is a particular problem. You want your students to plunge into it, focusing on its details so intently that they lose sight of everything else. Should you worry that they'll ignore fruitful analogies, or forget broad strategies, or miss altogether the key tactic of stretching the context? Generalization needs no fostering; we all seem to have a natural aptitude for it, even when we're consciously engrossed in particulars. It's a good idea, though, to step back with the students at times, to look at a process (like induction), or a principle ("look for patterns") that might work in other circumstances, or to note interesting features in the context. Even better, of course, should they come on these middleground or background issues themselves—but if you keep returning them to the problem at hand, not only will a line take shape for their digressions to weave around, but the natural impulse to generalize will, by opposition, grow more intense and will now have particulars to build on.

> Education is what survives when what has been learnt has been forgotten.
> —B. F. Skinner

Particulars: here are those invigorating instances we spoke of before, that the Ramanujan in us returns to again and again for refreshed insight and new wonder. The impulse to abstract always carries us away from the odd, the eccentric, the individual (after all, "idiot" derives from the Greek for "private person"). The beginning of work on a mathematical problem—in The Math Circle or in any context—is illuminated by this glory of the specific, which not only draws us in, but recurrently draws us on. Time for the timeless later.

You have to choose the topic for a course very carefully. Not any old problem will do, since the aim is to reward our students' efforts by having this work move them steadily and significantly beyond where they found themselves in the mathematical landscape. Their technique will have broadened, their understanding deepened, their sophistication grown. The question itself gives and demands that depth.

What we do, however, certainly can't be described as "acceleration", which is edu-speak for shoveling material into students as fast as they

edu-speak

de accelerate 160

can swallow it. First, works of art are to be savored rather than dashed past. Better an hour in front of a Monet than five museums in a day of Paris. The beauty, the meaning, the structure of a piece of mathematics unfolds only with reflection. And if—as we hope for all our students— each is to become more than a spectator, then it will take as long now as it did in Hippocrates' time to practice the craft and master the art. Second, to have people study outside of school what they are bound to meet within it can only guarantee future frustration. Our son spent his first grade abroad with us, and was fortunate to encounter, in Mrs. Midwinter, a superb teacher of mathematics in rural Dorset. When he returned the next year to his American private school, his teachers couldn't think what to do with him, since he had already mastered addition, subtraction, and multiplication—and change of base, thanks to pounds, shillings, and pence. Their unusual solution was to suggest that he take the year's workbook and go through it from back to front.

"Enrichment" is the usual antonym of "acceleration", but it fails to catch our spirit. We're not intensifying a standard mathematical diet by adding mental vitamins to it (tricks of the trade, show-off puzzles, glimpses of advanced results, a compendium of unexamined truths), but we're engaged in the wholly different enterprise of developing the architectural instinct. Making mathematicians, that is, rather than mathematical birdwatchers who compile lifetime lists of topics spotted. We inevitably encounter ideas fundamental and general enough to come up again in school—but with different emphases, and in a different context.

Once the students are engrossed in the problem, which fills the foreground of their view, we want to have them tease it apart, reassemble, and play with and around it—together. The point of this collegial play isn't only to subvert the age-old culture of school mathematics as yet one more outlet for adolescent rivalry but, through conversational exchange, to let diverse views differently reflect the whole: deepening the techniques, widening the metaphors, increasing the agility of each. This kind of camaraderie also knits together a web of other minds, which each one will be able to invoke: "At this point Sarah would say . . ."

Mathematics made collegially has even deeper advantages. Students new to The Math Circle expect its courses to be yet another forum for establishing a pecking order, but they forget about this when carried away by a sufficiently alluring problem. Early signs of this in a Math Circle class are a loss of all sense of shame: people so want to be in the discussion that they give up any false pretense of following it and will say: "I don't get this at all; could someone please explain it to me?" The almost instant reward is that, being in the conversation again, they can now contribute to it and even dispel someone else's confusion ("A mutual exchange of superiorities," in the Utilitarian formulation). A later sign is exuberant admiration for a classmate. "I'm used to being the bright kid, and when I walk in here I'm just stunned," as a high school senior said in a film made about The Math Circle by his tenth-grade colleague.* A roomful of egos becomes absorbed in *It* rather than *I*, and the surprising pleasures open up of forgetting yourself in the play.

Not only are these assembled egos moved into a conflict-free sphere (in Freud's charming phrase), but by having to give cogent explanations of what may still be half-formed ideas, each becomes conscious of exposition, as well as more fluent in the mathematical language they are jointly shaping—and this is a key part of absorbing the craft of mathematics. Moreover, the conversation makes each better at shaping critical remarks helpfully (since the work is shared) and better at letting critical comments abet rather than stifle creativity.

We've said that a Math Circle course is focused on and by a beckoning problem. You might think of that problem more structurally, however, as a feature in the mathematical landscape made prominent by how we are looking—or even as that whole landscape in miniature: a map or a model of it. For once again the context of our point of view begets what's seen from it, which develops that context further. Think, for example, of what happens in a course on Cantorian set theory. We ask naively, at the beginning, whether there are different sizes of infinity—and with nothing more than the notion of one-to-one correspondence (and proof by contradiction), an astounding world unrolls. Ingenious ways of making some correspondences, of course, come up—and even more ingenious ways of proving that others can't be made—but the variations that produce this rich tapestry are all on that single trope. Or again, so much of calculus is a development of the elegantly simple idea that ratios are preserved under diminution of scale—an idea launched, for instance, by the question: can we talk about the slope of a curved line?

Attention having been focused and play begun, we move together from the foreground of this landscape into its middle ground. The end now is to get the hang, simultaneously, of several things: the problem in its solution-space, of course; how mathematics is done (revealed in this as in each of its parts); and the contours of your own mathematical landscape, with its horizons that both bound and beckon. The way forward is always from this particular problem toward understanding the whole, but the view of it has been like those narrow old Japanese maps, centered on the traveler, that showed only the landmarks by the side of his road.

The students come to see the initial problem more broadly, and their own ways of looking broaden too. They develop styles of navigating that

*Noah Rosenblum, in Matt Paley's 2001 *The Math Circle*.

The basic idea of calculus differential

will work well elsewhere; their level of mathematical sophistication rises, so that they come to take as movable units of thought what before had statically filled their whole viewpoint. The point of view lifts, and the threshold of frustration with it. A lengthening history of little successes, and the companionship of fellow strugglers, help here too; equally important, though, is seeing from afar how many pathways there are in this landscape, where from close up, one had noticed so few. The closing down of one line of attack had then seemed disastrous; now one shuts only to open up another. The character of mathematical language changes with the growth of this middle distance seeing. Greater precision makes the mesh smaller, more (and farther-reaching) analogies make it more flexible. As a result, the net is better able to hold blunted approaches, undeveloped ideas, and alternative ways of formulating parts of the problem. The disorganized flurry of the more and less likely, which surrounds any enterprise, can drive you to distraction, but once well ordered, each insight can be put on hold indefinitely and accessed rapidly. It is this growing tolerance for putting things on hold that leaves the central arena of thought free to make new combinings. And within it, the span of attention lengthens.

At the same time, a background goal is also being approached: for we believe there is more in our remit than developing mathematical skill. We want our students to develop agile and probing minds for themselves; to be thoughtful before asking questions and fearless in pursuing the answers, with an imaginative intelligence. We want them to become skeptical about rigid rules or schemes of thinking, and simultaneously to become open to the suggestions of others. And so their performance rises to meet the gradually rising levels of their expectations and ours.

An effect (and then further cause) of this ascent is that they come to take over responsibility for the ongoing conversation—and hence for the development of their minds. Locke said that property is that with which you have commingled your labor, and if the work on the problem is theirs, and they know it, then the fruit of that work and the mathematical ground it grows from will rightly belong to them. Confidence and competence, as we pointed out in chapter 4, increase together. Now that they have some inkling of their relation to the mathematical world, where does it turn out they are? The answer always is—on the frontier. How can this be? It is always so from the standpoint of the explorer: he knows more than he did about the path he has followed—his particular problem and math in general—but with the x he sought still unfound, wild country remains before him. The frontier is everywhere. It is, like the amazing Sierpinski gasket, a space-filling curve. Perhaps this is because we are all still in our mathematical infancy; more likely, it comes from the nature of recursion. No sooner do you ask a question in mathematics than it suggests generalizations in all sorts of directions; no sooner do you answer it than the answer generates a spate of questions in its turn. In this midst we are always at the beginning, which is by turns exhilarating and exhausting. We are one with the Portuguese, whose motto before Columbus was "Ne Plus Ultra"—and after him: "Plus Ultra".

And now the solution to the problem begins to take shape, and the struggle to knit it into the known, via proof, is slowly won—won by fits and starts, really, as devices for proving are recast, refined, perfected; and after these twin victories, thought looks farther into the background. The issues are now how this problem fits into the whole (the theorem in the theory), and how the whole alters to absorb it—as space curves to accommodate the objects in it.

In this background, the emphases reverse again to what they first were: individual efforts are faired into a contemplation of the finished work, which gradually frees itself from its scaffolding of conjectures and personalities. Hence it is here that all those questions one couldn't help asking at the beginning—what's it all for, what's the point?—first become definitively answered. Since mathematics shows how things are— the hang of things—its least part carries the weight of explanation. Every part of it not only lends meaning to its neighbors but is likewise equally and unreasonably effective in how it bears on the world—and if we still wait to see the applications of some part, it is only because science has yet to catch up.

Beyond its thus pointing to how the world works, perhaps in the end mathematics *is* the point. Emily Dickinson wrote:

After great pain, a formal feeling comes.

More generally, after the long and strenuous wrestling with a problem, engrossment may suddenly broaden to a vista of relations that cease to arrow elsewhere and now round to a whole: you sense a form that deeply satisfies the architectural instinct. This is that a priori that Kant first detected; this is the revelation that dawns as invention, and brightens to discovery.

But this momentary glory soon fades, until what was astonishing becomes simply obvious (one thing revealed as another joins the network of what are no more than tautologies). Yet by being obvious, our understanding of how things *are* is heightened: we grasp what is no less than the tautology *that* they are.

This brief moment of formal appreciation may come at the end of the course, or not surface for days, weeks, months—when unexpectedly you think: so *that's* what it was all about! On moments such as these, the weight of compiled understanding balances that of accumulating open

The theorem in the theory

164

questions. We hear within us the clarion call from the seventeenth-century mathematician Vietà: "There is no problem that cannot be solved!"—and its echo in Hilbert, three centuries later: "In mathematics there is no *ignorabimus* [we shall never know]." Of course—a tautology again, since "mathematics" derives from the Greek "to know".

*What about Gödel's theorems*

## Beginnings

A piece of mathematics starts with givens and a question imposed on them, followed by means (such as deduction) for drawing out an answer. This happens in the context of similar questions, answers, and practiced styles of proceeding.

Students are our givens, to and on whom we pose a query. They bring their intuitions; the leader brings his skill in drawing out, to a context that develops afresh, through conversation, in each course.

### The Students

What happens depends on whom it's happening with. Although the great schemes to advance mankind assume that we are like enough for the same pod to protect us, in all that matters we seem to be stubbornly different. Since we take anyone who wants to come (that's important: they, not their parents, have to wish—for whatever reason—to be in The Math Circle), these differences can be extreme.

> It is important that students bring a certain ragamuffin, barefoot irreverence to their studies; they are not here to worship what is known, but to question it.—Jacob Bronowski

Each of us speaks a singular variant of our common language (choice of words, idioms, and constructions; different associations with and emphases on each), but language is usually so pliable a container of meaning that we readily understand one another. Our shakier relation to math makes such differences much more significant.

Four-year-old Grace had a beautiful new pair of shoes, with three artificial flowers on each. Her father heard her counting them: "One, two, two, three." "How many is that, Grace?" "Four." "Really? Try counting them again." "One, two, two, three." "Ah—you've counted the second flower twice, I think." "Yes I have." "Why is that?" "Because it's my favorite." As her father remarked, favorites, after all, should count for more. This is a far sounder conclusion than that reached by those child psychologists and educational theorists who (having never been children themselves) tell us that the mind wakes up in stages, with the Age of Reason being reached at some point in the seventeenth year. *HA!*

Of all the infinities that mathematics deals with, perhaps none is more awesome than this infinite variety in points of view—and none more wonderful than the infinite malleability of the mind. Like a stem cell, it grows to suit its context. "Youth is experimental," said Robert Louis Stevenson; once absorbed in a problem, you'll hear flying from every corner of the room the sorts of analogies and inspired inventions that we've anecdotally come to associate with vocation and genius. Just let tinkering loose.

We had been invited to give a Math Circle demonstration in a suburban public school's sixth grade. We worked with this class of diverse students on adding up the counting numbers—first from 1 to 10, which they calculated brutally, and then 1 to 20. Impatience with grueling sums drove them to hit on adding 1 to 20, noticing that all nested pairs amounted to 21. Also, there were 10 pairs of them, so all the counting numbers from 1 to 20 would equal $10 \times 21 = 210$. They quickly assumed that this would generalize for 1 to any n: you'd get $(n/2) \times (n + 1)$. But what if n were odd? A child of no previous mathematical distinction raised her hand: "Just start with 0 instead, and you'll have an even number of pairs, each adding up to n." Young Archimedes in Melrose, Massachusetts.

Assume at the start that no one knows anything—which helps encourage shameless asking and lucid answering of one another's questions. Still, the "nothing" we take them to know varies with the level of their mathematical sophistication. Age is the crudest of guides to these levels, so we sort at first by increments of a year for the very youngest, two years from around age eight, three from about eleven on. After a session or so we adjust and are ready to adjust again. An eleven-year-old girl ended up in a class of high-school seniors, and the mother of a seven-year-old found herself perfectly placed in her daughter's class, astonished by its succession of arithmetic revelations ("I was always told multiplying was just high-speed addition—so what are you adding, and how many times, in $1/2 \times 1/3$?").

A fair number of adults have joined our classes. The woman who delivered mail in the building where our classes were held couldn't resist what sounded like so much fun, and enrolled in a course with ten-year-olds that met when her shift finished. John, our barber, hearing that one of our courses was on the Pythagorean Theorem, asked if he could attend—it was something he'd always wanted to understand. "You won't mind that your classmates are about thirteen years old?" "What do I care?" he answered—and when we asked him, after a few weeks, how he liked it, he said: "You know what? I'm going to die a genius!" Most of our adult students fit unnoticeably (after the first few minutes of sidewise glances) into the class they stumble on: math absorbs minds equally. In one case a doctor, who was riveted by her eight-year-old's course, asked if we could run evening sessions for her and some friends, and this worked out very well (since they were all fairly knowledgeable, their course on number theory was differently geared to these different givens, but the style was pure Math Circle).

After the preliminary sorting by age, we aim for classes of at least five, ideally ten, and at most fifteen students. Diversity, again, dictates these limits. You need at least five people to take on the inevitable roles of doubter, conjecturer, exemplifier, prover, and critic (of which more later), and to make looking at a problem sufficiently stereoptical. More than fifteen and conversation begins to overlap and fracture, with no one sure of what was said by whom—including oneself. A collection, then, of a size to become a cohort.

Of course, most of these students have come from cohorts already, and inevitably bring classroom smarts with them. All but the very youngest are adept at getting along in school, where they have learned how to gain maximal results with minimal effort. They know how to read teachers (that telltale movement of hand toward blackboard, as the wanted answer begins to materialize), and how to collude with them against the anonymous Forces of Learning ("If you don't pester me with asking, I won't bother you with telling").

In a class we visited once, the teacher confronted a student: "I overheard you say you didn't feel you were learning much, Jeremy. But you know, you're learning a lot. It really is very remarkable, how much you've all learned this year. Don't you think so?"

"Yes, Mrs. Booster, now that I think about it, you're right." And so Jeremy took his first steps in irony, learning to make it the armature on which to build an accepted idiom in a formalized game.

It is heartening—another consequence of human malleability—how quickly students will exchange this mutual grooming for serious involvement with ideas, once they sense the changed tone. Their style of involvement with one another quickly changes too, despite the different presumptions they bring as to why math matters. Some are enthusiastic about its uses; some already love it for its music. A few want to defeat it, unhappy at the prospect of being weighed in its Olympian balance and found wanting. Some see it as a grappling hook to success. The self-aggrandizers come from the last two sorts, and it takes the most work to entice them away from competition and into the spirit of serious play. But all these motives tend to become redefined as the architectural instinct awakens.

The presumptions that students bring with them about what math actually *is* vary at least as much as their range of motives for indulging in it. Few have ever given a thought to what the foundations are from which mathematical structures rise—or even that there are such things as foundations. What are the Roman postulates (taken for the sake of the argument) or the Greek axioms (statements worthy of belief) compared to these dreamers' certainties of the way the world just is? Why prove what we know? How prove what we don't?

What about gender differences, when it comes to aptness and avidity for math, and proficiency at it? We've seen countless examples of *beliefs* that there are such differences (with girls always coming out on the short end) but precious little to support them—beyond the undoubted effect they have as cultural *bien entendus*.

We've found that girls tend to lead in Math Circle classes up to about age twelve, then back off, and their numbers begin to decline with the onset of adolescence. We get the sense that parents view math as more important for boys, and that girls hear from their friends that it isn't cool to be too prominent in a traditionally male realm; the competitive aura of math in school is again a boy thing. Besides, the slower maturity rate of boys prolongs in math an atmosphere that *they* may delight in as nerdish but girls put away as simply childish. Earlier emotional maturity allows girls to prosper in history, literature, and languages, so why should they fight out a rear-guard action here?

Perhaps different chemical tunings of male and female brains play a role too, if indeed the former by nature incline more toward the mechanical than the latter—but only if math is presented *as mechanical*, robbed of what we've been calling the Ramanujan, as distinct from the Hilbert, component: the inspiring and renewing savor of the singular, without which abstraction is hollow. And if it is true that stress leads to a rise in serotonin production in the male but a drop in the female brain, with consequent elation or depression, then the traditional school association of math with competition will strengthen its ties to the masculine world, and shadow even the noncompetitive conversational style of our Math Circle.*

These are both current "ifs", their putative existence buttressed by hypotheses about our hunter-gatherer origins. But even were these inclinations established, it is vital to keep in mind that the conclusions would follow from narrow statistical differences, and that what holds "by and large" is irrelevant to the particular people sitting in the room with you—who, in the case of The Math Circle, have chosen to be there. Here and everywhere, nature is in constant parley with habit. One mother wrote to us: "My daughter was accepted into Boston Latin School yesterday. She ranked first in math out of close to 2,000 sixth-grade children who had taken the exam. I really have to thank you for having turned her from an 'I hate math' little girl into this young woman very much into math."

---

*Our information comes from an article by Larry Cahill, "His Brain, Her Brain," in the May 2005 *Scientific American*.

Scientific findings, whatever their worth, become part of the cultural context. For so long and in so many subtle ways, ours has reinforced the belief that girls just are not fitted for these rigors. So when adolescent boys and girls get a math problem wrong, boys tend to say that there must be a misprint in the text, but girls say that they have now reached the limits of their ability. We knew a girl who, in a precalculus course, found her text asserting that $1 \times 1 < 1$. She hurled the book away, in despair at the depths of her ignorance. Her boyfriend picked it up and—correctly, for once—said this must be a typo for $|x| < 1$.

All these differences, along with those of susceptibility (as for the algebraic or geometric) and of liking some things (proofs by contradiction) and abominating others (inequalities) will enrich the flavor as the broth simmers. Its savor only fades when parental ambition intrudes. Doing well in math is, unfortunately, a ticket to advancement, making an already prickly subject bristle just that much more with anxiety. This is one of the reasons we're quick to disclaim any advantage that might be gained from coming to The Math Circle. For all that parents know what's best for their offspring (especially because they can more easily see windows of opportunity falling shut), we've yet to find an example of a parent's imperatives successfully conquering a child's wish to be elsewhere. We remember only too well the young Russian whose after-school hours were gridded with ballet and skating classes, piano and swimming lessons, riding competitions and The Math Circle—and an appointment at day's end with a nervous breakdown. Mathematics is freedom, and no respecter of any, save inner, compulsions.

There are certainly those whose inner life has so developed as to make The Math Circle unsuitable (all the more reason for starting very young). Serena was convinced that there is a fixed body of truths; it was the teacher's duty to impart, and hers to receive them. Only her love of arguing, in the end, argued her out of this conviction. Simon saw himself as a soldier in a life of trench warfare; he would give name, rank, and serial number—no further commitment—to any enterprise. Thomas was passionate yet passive; the truth meant so much to him that he knew he had no right to meddle with it. Perhaps love, adventure, or adversity would later make an approach such as ours appealing.

> If a man's wit be wandering, let him study mathematics; for in demonstrations, if his wit be called away never so little, he must begin again.
> —Francis Bacon

## The Leaders

Math teaching is seen as an easy job—questions have a single answer, and it's in the Teacher's Guide. All you have to do is explain the rules clearly. But leading a Math Circle class is very different. You not only have to know your stuff, you must be unwilling to tell it. More precisely, you need to know the focal problem and its context well enough to help your students navigate through it; you need to want *them* to do the exploring, with as little help from you as possible (unlike the fellow who taught lateral thinking by showing his audience how *he* solved the problems he had just stated). In short, as we said at the outset, you have to be a Sherpa.

*ironic*

A Sherpa with a difference, however. It is up to you to choose the mountain they will climb: a topic you not only know well but love. For you need to communicate an excitement that will generalize from this corner of the landscape to math as a whole—math *per se*—and from you to them—math *per eos*. This is an underlying duality in all of teaching: what happens with the subject is shadowed in those whose subject it becomes. A student, who thought the whole Math Circle curriculum cycled through the courses he had taken, once asked us how we could stand teaching the same thing year after year—not having yet appreciated that math is endlessly deep and people endlessly different. Fun and the fungible are poles apart.

The topic may be one you've recently mastered or have been mulling over for years. In either case, you'll know where to lead the students in it, and later, how to follow their leads and make helpful adjustments to them, with new prospects ahead for all; since everywhere is frontier at this (and perhaps every) moment in the evolution of mathematics, unexpected vistas continually open in even the most well-traveled landscape. Each city here is Dr. Johnson's London: when you grow tired of it, you have grown tired of life.

In choosing a topic, keep in mind the point we made in chapter 5, when speaking of symbols: means to an end seem to be more readily absorbed than the end itself (another aspect of teaching by indirection). So if you want your students to learn some topic A, find another topic, B, which A entails—and have them work on B. You'll see that A is effortlessly taken in stride (although B itself may escape).

Let's say that for some reason or other you thought it important for your young students to learn their times tables (whose contents these days seem to be called "number facts", which puts them in a class with what you need for trivia quizzes on Friday nights). Set them instead to puzzling over which of the first hundred or so integers are prime. This is always a heady pursuit for our pattern-hungry minds, and in the course of it such "facts" as that 51 is (disappointingly) $17 \times 3$ are hit on—and stick. Once they dope out how the times tables are made ("It's just adding, and once you know $3 \times 4$ you get $4 \times 3$ for free"), a very few significant encounters tamp down the actual values ("and it's 12, like feet and dozens"). You could also have had them figure out how to add fractions: finding a common denominator is tense and rewarding, and puts times tables in their right (ancillary) place. Or if you want them to master deductive reasoning, put in front of them an intriguing topic in Euclid. If the subject is calculus, embody it in a particular physical problem. It takes some thought to find a good B topic for each A—fortunately no system of facile entailments is ready at hand.

Now that you have your goal and an indirection to it, you need to learn how to point without waggling your finger. This is an acquired skill. It begins with asking a question that the students will both want and have to sharpen ("What happens if we add to the Euclidean plane the points where parallels meet?" "How many roots does a quadratic have?"), and continues by answering their questions with what will turn out to be further (perhaps more focused) questions. You need to have a trajectory in mind, planned in consultation with the topography of the problem, but no more worked out than that of the sentence you're in the midst of saying. Should you see a discussion heading for a cliff, you have to decide if this is a plunge worth taking. As in mathematics generally, intuition guides your exploring of a formal terrain.

That course about interesting points in triangles (mentioned in chapter 4) began with the students thinking about the perpendicular bisectors of two sides. As the leader was drawing these on the board it occurred to him that he might curve them to make it appear that they were parallel. The children had not yet had a course in geometry, but brought with them a well-developed Euclidean intuition. Let the lines intersect and the course could go on to discovering the circumcenter; curve them, and hours of chaos, clarifying at last to some Euclidean postulates, lay ahead. It was a split-second decision. He weighed up their personalities, energy, enthusiasm . . . and curved the lines. While the course never got to the nine-point circle he had seen as its goal, it led through passionate discussions about physical and mental diagrams, and the role of assumptions, to ownership of this bit of land in Euclid's domain.

The initial question helps to gather diverse energies together, but for its propulsive force to continue driving the conversation on, you need to talk with, not at, your fellow voyagers. This means not talking up toward the mathematics ("Let the floor function now be defined as follows") nor down to charity cases ("You don't see that? Well then, let me try to explain it to you in another way"), but speaking as you speak to yourself. A very sound piece of advice, however, that we were once given by a Montessori teacher, was to talk to a child as you would to a fellow adult, but the moment you see eyes glaze, stop dead: nothing is gained by forcing on a one-way discourse.

Your way of speaking not only establishes the terms you will be on with one another and with the math, but the quality of the whole undertaking. Some say that the smile or frown on the face we first see shapes our lives; it certainly seems to shape the life of a class: the tone set by the leader's initial response is the tone that tends to persist. You want your students to sense your interest not only in them but in the math; they should know from the start that everything is open to inquiry, to constructive doubt, and to wild conjecture; that laughing will be with, not at—for this is to be an enterprise of high rather than low comedy. You therefore need to dissociate yourself from the Disneyesque belief we spoke of before: kids are fundamentally clowns and prefer clowning, and if we don't jig around saying "Wow!" sawing the air with our hands and popping our eyes, they'll sleep. Such antics may in fact make them wonder if adulthood is all it's cracked up to be.

You need to place yourself equally far from the all-too-familiar type at the other extreme, who recites the *Black Book of Sea Law* before a single ship has set sail. These are the people drawn to mathematics because it is rigorous—and conclude that it gives them license to be rigorous too. If you fear that your students are beasts about to break loose, let the sheer, intense fun of the undertaking soothe your breast and theirs. If the conversation is beginning to gyrate away from its focus, either reshape the topic around the focus newly framed, or deflect the conversation back as unobtrusively as you can. You're unlikely to try embarrassing students, wanting them instead to feel comfortable with making the daftest suggestions, but you may be inclined toward too lavish praise, which can make people equally anxious. Your aim is to create an atmosphere of intellectual intensity, emotional ease, and contagious enthusiasm. This is the atmosphere in which weighty problems begin to rise, as if by their own volition—but in fact through the collective buoyancy.

There are, undeniably, spirits of gravity to work against as well, for the ductility that binds a class together can just as easily disperse it. Crowds (and five may well be a crowd), especially of children, are game for anything, and everyone's sensors are atingle to detect what the mood is to be. We all know that mocking sing-song, at least as old as Roman schooldays: it waits in the wings, as does fervid attention. Between these two are all sorts of diverse inclinations and reservations; soon enough they settle into a consensus, and the spirit of the class, whatever it is, will then seem to have been inevitable. It depends wholly on the leader to set the style from the first moment—not by pronouncements and pleas but by those off-hand signals that say: "this is the call to adventure, and this is how we will carry ourselves in answering it."

To those scholars coming from the semi-monasticism of mathematics departments, such skills of generalship are often unknown. They have been used to that peculiar give-and-take with the world that consists as much in quietly listening to what phenomena tell them as in imposing their inventions on it; their conversations largely pick up and trail off in

172

conjecture; they know that evasive ideas call for cautious approaches. Who are they, so prone to thinking "If . . .", to rally troops with an unequivocal "then"?

We had taken on one such colleague eager to teach in The Math Circle and left him on his own with a roomful of eight-year-olds. Ten minutes later a parent came up to where we were teaching: "You'd better go downstairs, there's someone out in the corridor, crying." Visions of assault and suits for child molestation. But in the corridor downstairs it was our colleague we found in tears. "I've never said no to anyone," he told us, "and I'm not going to start now." In his classroom the paper airplanes glided serenely over the tumbled chairs.

It isn't hopeless. Leaders are as malleable as students and can work themselves up, for the greater good, to put on a Prince Hal disposition, though they feel none—and here too, the expectation can produce its fulfillment. So much of human engagement rests on responses reinforcing one another and creating a gradient, that initiatives can become intermediate habit, and habit second nature.

Their charming shyness makes many mathematicians eager to communicate mind to mind but uneasy at looking eye to eye. Eye contact isn't necessary, but it does help. It certainly helps to know your students' names. Both are passageways into where they are in the evolving conversation: what parts of the picture they see; the drift of their understanding; how clear they are about the groundswell of strategy and the ripples of tactic, and about what has been established (and how). You want to be able to look from their selves at yours, the others, and the math.

This focused empathy is a great support when it comes to dealing with the student out to dominate or scuttle the enterprise. Cajoling him now, talking alone with him later, are both made easier by having a sense of his motives. A needy sort, who can never get enough recognition? Frustration? A lack of engagement? A bad hair day? If you see the course from his standpoint, you can more readily decide if the mutual benefits have dwindled to zero.

A triumphant (and at times barbed) cry of "It's obvious!" can destroy the collegial enterprise—and not only for those to whom it is anything but obvious. Ask someone who so calls out to explain it to you and the class. This may do some good for their humanity as well as for their powers of exposition (and may uncover a corner or two that wasn't so obvious after all). It gives an opportunity as well for a conversation about the peculiar synthetic a priori quality of math, that having invented our way to an insight, we now can't imagine how we could ever not have known it.

What if the discussion has somehow developed maimed, or misunderstood, or misleading premises? Indirections are best: "You mean . . ."

and produce an example that brings this misapprehension to light. A class of seven-year-olds saw the rationals as spaced out in a row, like rippled points, with minute gaps between any two. The leader agreed enthusiastically, then asked them to pick an adjacent pair and had them construct its average. "Now where shall we fit this in?" Once the horror of rational density took hold, and they agreed in awe that these filled up the number line, should she have brought up $\sqrt{2}$ ?

Even the wildest suggestion must be allowed to run its course; more is often learned from following a promising route to a dead end than from clicking smoothly over the points to the gleaming station. Greeting a conjecture with a smile or a frown as much undermines self-reliance as does moving your hand in anticipation toward the board. Mathematics is freedom only when the constraint of a safety net is taken away—so long as falling is taken in the same spirit as somersaulting onto the trapeze.

You'll overcome the temptation to mistake the role of guide for judge when you think the consequences through: you're trying to prove a theorem, not yourself. The only authority to be invoked is the structural integrity of the construction. And if in frustration a student (like the Russian seven-year-old we described in chapter 4) says, "Just tell us the answer!" simply explain that having it told, or looking it up online, is as valuable as having someone else do your exercising, since understanding the problem is an end, but also a means to the greater good of tinkering your way toward mastery of this craft. A lot of teaching is acting, and if you take on an air of shared ignorance, your students will suspend their disbelief. Not only will you better sympathize with their perplexities, but you'll be in a state yourself to catch the whiff of unresolved issues.

As in every kind of play, timing is all, and the hardest thing for a leader to learn is when to let the problem keep tugging everyone on, when to slow down in order to pick up an important piece that fell off, when to recoil, when to charge—to go on or let it go. If you step in with a hint to relieve the frustration of five-year-olds trying to figure out how to add 1/3 and 1/2, you will win their gratitude but lose them the risky chance of revelation. Read the eyes for glaze and the minds for mood; think your way into the least and most confident there—and then guess. Keep suggestions for other approaches on tap (Should we try to visualize this? Is it time to call algebra to our aid? Are we thinking too narrowly? Too broadly? Would it help to say explicitly what the obstacle is?). Sometimes the best thing is to put the problem temporarily on hold with a historical or biographical digression, or with what might be a covertly relevant game (such as function machines)—something to enrich the context and at the same time let people catch a second wind.

Choice, risk, leading, following, timing: a Math Circle class is a high-wire act, and the best classes are always (like the best sessions of math in your mind) on the verge of chaos, which is what gives them their intensity. Here again the evolving cohort is an ally: the interweave of their ideas and personalities is the real safety net, and it depends very much on the leader to turn what too easily leans toward a fellowship of cynicism, or the camaraderie of clothes, into a generous admiration of one another's insights.

Of the second-rate ruler, people speak respectfully, saying, "He has done this, he has done that." Of the first-rate rulers they do not say this. They say: "We have done it all ourselves."
—Lao Tze

What's surprising is how quickly this can happen. We mentioned earlier the demonstration of The Math Circle approach we once gave to a London audience of more than a hundred and fifty nonmathematical adults (and a smattering of children). We were met with audience passivity, tempered by nervousness, skepticism, large hats, and crutches. Within ten minutes, however, people had plunged into the problem, and suggestions for dealing with it came from all over the auditorium. A reporter from the *Financial Times*, writing about it the next day, said: "It is most odd, but a bonding thing seemed to be happening." Testimony, really, to the rousing of the architectural instinct, that draws strangers together on common grounds. A Quaker meeting may take years to "mature" (for the Spirit in the Midst, that is, to speak through the voice of anyone present); the spirit of hidden structure, clamoring to be heard, can mature a Math Circle meeting just like that.

Clearly so much depends on the leader's personality; yet there are no types, no traits, that are best, since the freedom, which is mathematics, must extend here too. Whatever we recommend is to be taken (as a mathematician would put it) modulo each individual: in the terms and within the interpretations of his world. Just as we singly understand mathematics with different emphases and from different angles, so we show our enthusiasm, we plan how to let our students discover or create, we *connect* with them, differently. To legislate this would be to make the organic brittle.

Whatever one's personality, it acts at the beginning of a course as the conduit from the mathematics to the students: the abstractions are realized in the leader's character. As the course goes on, however, not only does the centroid shift from leader to students but from personality itself to mathematics. The scaffoldings of I and You fall away to leave the It we have been working on revealed to a Buddha-like eye. This happens in stages. The supposed power position of leader smoothly morphs into that of accompanying guide. Math Circle leaders have come from the ranks of school teachers, professors, post docs, graduate and undergraduate students of math; from people in physics and the computer sciences;

from humanists who are true amateurs of math; and in one case, a high school junior who had been a Math Circle student from the time he was ten (see Sam's thoughts on his experience in the appendix). Where will they continue to come from?

It takes actually participating in Math Circle classes fully to breathe their air of serious play. Surely, then, former students are ideal candidates for future leaders. But as Math Circles develop far from their origin, their leaders will profit from sessions with us: taking part in classes as students, trying out their style on one another, and then on the students they have just been sitting among. And then it's up to them, on shakedown cruises with their crew of students—reckoning up and on one another's strengths, weaknesses, and inclinations; finding out how personality and opportunity mesh; and inventing a common language, as The Math Circle speciates in unforeseen ways.

**The Math**

We said that not any problem will do. Why not? Why not pose, for example, an open question, in hopes that the communal enthusiasm and collective naiveté will lead someone to stumble on an answer? This was the aim, as the story goes, of a mathematician giving an advanced course to excellent students at MIT. The final exam consisted of only four problems. He left them to it, and when he came back an hour later, to ask how it was going, saw only gloom; they knew how good they were, yet none had gotten anywhere with any of the questions. When he returned two hours later, one was in tears, another had pulled out great tufts of his hair, the tang of panic was in the air and heaps of paper on the floor. "Any luck? Anyone? No?" He then told them that these were all unsolved problems—and barely escaped with his life.

Perhaps the story is apocryphal. One current program, however, has been described to us as posing questions far too difficult for the students to solve, so as to teach them to break problems down into manageable subproblems. This is an interesting strategy—so long as the students don't break down first.

A Math Circle leader having settled on a compelling question, will modify it so that it needs no more mathematical background to understand than his students would naturally have: a problem appropriate for a universal audience. You don't want a question that forces them to trundle up the siege-guns of calculus, or topology, or group theory— unless they all have these standing ready from previous courses. You don't want to have to do the crucial work for them, telling them things they aren't equipped to discover. These eliminations still leave an enormous number of questions, none of them marginal, in geometry, algebra, number theory, combinatorics (the favorite hunting ground of contest makers), logic, and set theory, for a start—and if you find that you really have to maximize a function, say, classical formulations can be rediscovered and intuition developed to fit local circumstances.

Every problem is the join, tip to tip, of two infinite cones: one opening upward toward conjectures, the other down to foundations. The particular problem from whose seed so much of the subject in particular, mathematics in general, can grow, should bring with it an unexpected view of the familiar, or suggest a structure where none had been thought of (More than one infinity? Numbers whose squares are negative? Seven dimensions?) But it must seem just an outstretched finger away. Far enough fetched, then, to be surprising—even unsettling; not so far that you fear your mother-wit can't solve it. *Accessible mysteries:* delectable opportunities for your preparation to let itself loose on.

*Tell Eva*

- If you're bouncing a ball off a wall to your friend, where should you make it hit so that it gets to him? What if he now moves closer to or farther from the wall than you? What if you want to bounce it to him, or back to yourself, off two angled walls?
- What numbers form Pythagorean triples?
- Is there anything the same about different polyhedra?
- Can we tell one knot from another?
- What numbers are both triangular and square?
- What's the shortest path made up of straight-line segments inside a triangle, if it touches each side once?  *Good question.*
- Can you trisect an angle?
- Can you find a function whose graph passes through two given points? Three? Seven? n?
- Can we tell if this series converges, and what it converges to? What about other series? What about general tests of convergence?
- What is $i$?  *Yeah!? What is it?*
- Can a visual proof be rigorous?
- How many different 17th roots of 1 are there? What do they look like on the complex plane?
- What maps can be colored with only two colors? Three?
- Is any natural number the sum of two squares?
- Can you share out twelve cards among 3 friends? Among 4? 5?

The list is literally boundless: it only takes looking for a loose end dangling down from a tangled structure, which tugging at will unravel. As an exercise, write down for yourself the beginnings of a list of intriguing problems about nothing more than triangles: there's the world in a grain of sand.

These accessible mysteries can hardly fail to arouse imagination, the sense of adventure, and the architectural instinct. At first it depends on the leader to keep the broader, background, concerns alive; but soon enough the variety of voices in the room will move the conversation up or down a level from the immediate concern.

Ideally the problem will not only be solved within the ten-meeting span of the course, raising the threshold of frustration and the lintel of confidence, but will open up related problems to carry mind and the craft forward. Much of the impetus for this comes from work on the problem itself, and the gambits that failed as much as those that succeeded. Part, however, will come—as we've mentioned already—from the looseness with which the problem was first posed. That imprecision not only develops the crucial skills needed to make saying and meaning converge but unlocks the possibilities in the nature of things.

"Are there as many even numbers as counting numbers?"

"Sure—they both go on forever."

"Does that show there are as many?"

"Yes, the number of each is infinity."

"Is infinity a number?"

"Well, isn't it?"

"Maybe it is. When you add 7 and 5, what do you get?"

"12."

"So the sum is different from either of the numbers being added. What's 7 plus infinity?"

"It's infinity—but that's just like zero: 0 + 7 is 7; and zero's a number."

"You're right. Can you tell without counting which there are more of in this room, people or chairs?"

"Chairs—because everyone's sitting down, but there are chairs left over."

"And if all the infinite number of chairs in a gigantic room had the counting numbers successively stuck on them, could all of the evens then sit down?"

Such questions unlimber imagination and swivel it toward what hadn't been thought through. If you present the problem to the group, however, with all its ambiguities removed, you not only take away from the student any chance to see the hidden ratchets and levers, but also confront him with an arbitrary-seeming set of constraints, rather than handing him a diamond to turn as he will. Think again of the finicking grammar in the $\varepsilon$–$\delta$ definition of limit, which we struggled to perfect in chapter 4:

The function f(x) approaches the limit $l$ as x approaches a,
if and only if for all $\varepsilon > 0$ there is a $\delta > 0$ such that $0 < |x-a| < \delta$
$$\Rightarrow |f(x) - f(l)| < \varepsilon.$$

How many students and teachers would have trouble picking out the one right formulation in a multiple-choice test? It's not only dizzyingly formal, it's also unintuitive: why shouldn't the quantifiers be reversed? Why strict inequalities? Why doesn't the implication run the other, or both, ways? But if you ask a student to give you instructions for making a function's output approach a number *l* as its input approaches a number *a*, an urge to mill the parts neatly together will produce a vivid mental image and a lively piece of language, rather than a memorized formalism.

A course—especially for the youngest students—may begin not with the problem it will come to center around but with some preliminary questions meant to dissipate the shades of the prison house that lengthen so quickly, as we try to anticipate demands on us and just get along. Think of the questions the farmer-turned-teacher, Mathieu, asked his Rousseauvian children in that wonderful countercultural film of the 1970s, *Jonah, Who Will Be 25 in the Year 2000*:

- Does the wind feel the clouds?
- Does the bicycle know it moves?
- Does water feel anything—even if it boils?
- Where does the word "sun" come from?
- Does it know it's called the sun?
- When we move, does the moon move too? And if we stop, or change directions?

Our stretching exercises are in rather the same spirit:

- What's the last number?
- Three people are in a room and seven go out. How many have to go back in before the room is empty?    4
- What is 4 lions times 3? What is 4 lions times 3 lions?
- What is 2 plus nothing? What is 2 plus something?
- (To seven-year-olds): How old were you six . . . seven . . . eight . . . years ago? [Think about this: to what "you" do these questions refer?]
- (To thirteen-year-olds): What's the biggest idea in math you understand? What's the coolest you don't?

Questions like these not only shake a sleeping imagination awake but tell you something about the student's mathematical whereabouts. "What's two plus something?" "Three!" Ah—numbers are still adjectives for him, not yet nouns on their own (or he is still a stranger to fractions).

No need to confine these questions to the beginning of a course. Asking them opportunistically can capitalize on the moment. Five-year-old

Max was taking the seeds out of a pomegranate. "How many are there?" "53." "More like 5300!" said his babysitter. Max picked one up and popped it into his mouth. "Minus one," he said. A humorist, and a born diplomat.

The best Math Circle problems aren't just open-ended but open-middled: full of junctions for switching to destinations other than the intended, if it takes too long to get there or the students carry the course down a siding worth following. Let's call these "bristle points". Take, for example, the course for young adolescents on "Interesting Points in Triangles" (mentioned in chapter 4 and described in chapter 5). You'd like them to end up discovering the Euler Line and the nine-point circle—but just proving at the beginning that the perpendicular bisectors concur is a significant triumph: they learn that mere transitivity is a powerful tool. Going on to find the angle bisectors concurrent invokes more of Euclidean geometry, and concurrency of the medians even more. To then find a triangle's orthocenter, by seeing it as the circumcenter of a circumscribing triangle, is to glimpse the beauty that now not Euclid alone saw bare. Any of these way stations serves as a satisfactory conclusion. An equally appealing course, however, on Eulerian circuits (which maps will let you make a tour going on each road once and only once?), really has to have its question answered, but many directions (some to still unsolved problems) then open from it.

Is this the only possible, or the best, road we are following? Not at all; despite the multitude of curriculum developers, a royal road to mathematics is still unbuilt. Not only have different problems more and less vulnerable points of entry, but different people are differently susceptible. Some like to begin with a puzzling detail (like the child in the myth who tugged at what he thought was a buried ring—but which turned out to be the curled-up tail of the dragon who girdled the earth). Others prefer to work from whole to part, while yet others like dashing back and forth between these extremes. Some people need to have concrete examples at hand, some like an abstract music. Hence we cannot but move simultaneously along a network of paths, these having enticed some students, those others—and these paths not only cross time and again, but direct imagination and converge on the problem, as it is newly understood and re-expressed.

## Connecting

There is the goal and here we are—but it's not Euclidean geometry: these two points don't determine a straight line. Why not? It isn't just that we're made of such twisted wood (as Kant pointed out). The thing about discovering is that you might not. All those various pushes from inside each of us and from each with and against the others; all those tugs from the goal, making now this route, now that, seem the most promising; all those bristle points, all the will-o'-the-wisps along the way: these lay down switchback paths, skirting around bogs and quicksand, so that it rightly seems a miracle when we get there.

What evolves in this journey, punctuated with pauses to pick up your bearings, steps back and leaps forward, are the relations among the explorers, and within the math, and between the two. Those who were nervous to start with become eager, and then engrossed. They blend into a cohort, but in the end look out *from* the math rather than collectively or singly into it. Hunting for clues and gathering data turns into hypothesizing, then to proving, and last, to reflecting—as the mathematical structure coalesces and becomes at the same time abstract and vivid.

**Intuition Grows**

Before we come to grips with a problem, we grope at it. Since the pressure our minds and the problem put on each other will shape the course of inquiry, any first contact will do, so long as it adheres.

When, for example, we work on Eulerian circuits with eight-year-olds, you'd expect us to start roughly like this: "We have a friend who loves to travel. He doesn't mind going to the same city again and again, but he's very fussy about roads: he wants to go on every road his map shows, but only once on each. Which maps will let him do this, which ones won't?"

In fact we begin somewhat differently: "You have a friend who loves to travel—oh, what's his name going to be?"

This may seem distracting, or at best inconsequential, but in a way it is the most important question we'll have asked, because it brings the students at once into a conversation where nothing is at stake, establishes a light and complicit tone, and makes concrete what will so rapidly become abstract. The power of imagination now propels not some but *our* traveler along his ever more baroque routes.

As in each of our courses, the students make wild stabs at an answer (which surely must be waiting to be found) and come up with complicated examples and counterexamples.

"It works on any map!"

"Not if the map has 10,000 roads!"

"It might, but you'd never know."

Frustration suggests questioning the question, and one discovers that some rethinkings illuminate and others destroy: "Why doesn't he take an airplane?"

They quickly come of themselves to see that the simpler examples are the more telling: just as in trying to guess what another's function machine is, the fun of giving gigantic inputs soon changes to cannily choosing inputs like 0 and 1.

What happens next happens in every course: they hit on the uses of abstraction and begin to draw their maps without landscape features, with points for cities and lines for roads—and now thoroughly launched in this mode, see that wiggly roads hinder more than help understanding. Someone comes up with the simplest map that works:

And soon after, the simplest that doesn't:

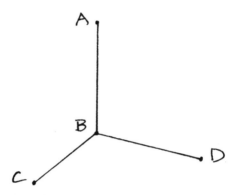

The inevitable joker will produce the even simpler map,

$$A \bullet$$

which may eventually turn out to be a joke that leads toward rather than away from the undertaking.

They have already begun to forget their egos in this engrossment with the question, although eagerness to be the first to solve the riddle is certainly still a driving force. Soon, however, the network of examples and counterexamples, of hypotheses that failed but might still be rescued and those that continue to look good, will so fill their memories and absorb their attention that the sense of self will all but disappear, just as the persona of the traveler has been replaced by the structure he moves through.

This is a time of concocting examples and collecting observations—though as Darwin remarked, you can hardly observe without an hypothesis. Traits of character weigh in here: some hug their hypotheses to themselves and produce example after example to confirm it; others are skeptical and stake out their territory as a search for exceptions to any suggested rule. Some are eager to endorse all candidates in the ascendant, some relish the role of naysayer. Attractive generalizations, unacknowledged assumptions, and seductive errors tangle the web. The conjectures, however, begin to grow less wild; instead of springing up in vacant territory, they tend now to feather out from the most traveled lines of previous conjecture. This may lead down worm (or even black) holes or colonize empty squares in cross-hatching thought.

So much data and meta-data comes to accumulate that of itself it brings out the urge to classify—which, while it facilitates thinking, entails hidden hypotheses all the more powerful for being covert.

This first, frisky stage is full of the pleasure of letting imagination and tinkering loose. Baudelaire knew this:

> For the child, in love with maps and prints,
> The universe is equal to his vast appetite.

This fun bonds together the travelers and the world they explore. It also stocks the cellars of collegiality and optimism against the days when the fog closes in. What has been fundamentally happening in this stage is sizing up the problem: it reminds me of something else, which I understand; here's an approach which seems to fit better than that; were only such-and-such true, then that would follow; if this were so, then the situation would be of this sort, or that over there would be an instance of it . . . 'tis very like a camel.

In fencing you try to establish a measure: the distance that will let you safely strike through your opponent's defenses, and that is what we seek to establish here: the right remove from, the right angle toward, the problem—taking its measure. Perhaps we err in thinking of intuition as a faculty or capacity in isolation; instead, it may always be a matter of *intuition for:* suiting ourselves to *this* problem's character.

The next stage grows out of the frustrations in the first. Once we feel we have listened to the problem long enough, we start to impose patterns on it actively—for regularity must lie under its irritating gaps and surprises. Notice that this "must" isn't an imperative of the world, but of our thinking: our architectural instinct makes patterning beings of us all. However fruitful the diversity in temperaments may be, this imperative will unite extremes, from the brash to the cautious.

How does it work? The fragile hypotheses that let us observe in the first place begin to gather importance and edge observations out, as the weight of data exceeds our capacity to put it on hold. To our great relief, the hypotheses now coalesce into a much more tenable structure, which we preserve for as long as we can by turning a blind eye toward exceptions or interpolating second-order hypotheses that will smooth and simplify away the blemishes of detail. Abstract recursion bears its share of the blame here too.

Look again at that problem we described beginning on page 71 in chapter 4: what's the most number of regions into which chords drawn between pairs of n points on a circle's perimeter will divide its interior? You recall that 2 points gave 2 regions, 3 points 4, 4 points 8, and 5 points 16—so that $2^{n-1}$ regions was the obvious answer. Yet draw it how you will, 6 points generated at most 31 regions. After trying again and again, an earnest student asked us: "How long will it take for the pattern to re-establish itself?" From childhood induction on ("Anything I say three times is true"), we cling to patterns as closely as did the ancient Greek mariners to their shores.

Everything gets fuzzy during this time when insight no more than stirs in its sleep. Trying to preserve a pattern against the evidence can be tiring, and with fatigue the problem goes in and out of focus, as do our most recently acquired skills. Young children will think: since a(b + c) = ab + ac, it must be that a(bc) = ab·ac. Older ones will be uncertain whether we'd agreed that for all ε > 0 there is a δ > 0, or was it that there is a δ > 0 such that for all ε > 0, or one of the vice versas, and does it really matter? Everyone falls back on the more reliable older truths. To add 1/2 + 1/3, find a number that both 2 and 3 "go into". Ah, 5—since 2 + 3 = 5, and addition is much more trustworthy than this newcomer, multiplication.

The leader plays several very important roles at this point. His quiet optimism keeps the climbers roped together. He may see a chasm ahead that he will want them to skirt or step over: these gaps will still be there to explore the next time around (some quadratics have no roots, some two, some a double root; let's leave it at that, without a wink wink or a nudge nudge). With everyone's thoughts darting off in different directions, he also serves as their centroid of memory, and may urge following a suggestion of five minutes ago—but whichever route is taken, he preserves the context. If a line of thought is suspended, he will be able to reel it in and connect it up again to conjecture and consequence.

The Sherpa's most active role, however, is giving students the equipment they need for the steepest ascents. Ideally, of course, they will produce it all themselves (in a deep sense, mathematicians have to reinvent all of their wheels)—but this is often too much to hope for in the ten hours of a Math Circle course. Take that problem of tiling a rectangle with incongruent squares. They have probed it to a standstill and there they are, in their frustration. The leader sees that the odds are enormously against anyone thinking of tiling with *rectangles* and then treating these algebraically as if they were squares (with areas $x^2$, $y^2$, and so somehow on). She therefore suggests a strategic retreat: "We're not getting anywhere with squares—can we tile it with rectangles?" She subsides, like the dormouse, to emerge if they need a further suggestion. As they pick up these scattered clues and follow them to the goal, they will have handled them so thoroughly as to make them all but their own.

Only
10 hours
?

This benign plagiarism conforms to the intellectual property law of mathematics: the proof, the technique, the conjecture may bear Gauss's name, or Euler's, or Lagrange's, but it belongs to Mind.

This is a long haul, but during it unsound leads will drop away, and what seemed at the time like diversions from the straight path may turn out to be short-cuts—or at least to have added suggestive background detail (that the problem came from a populated locale may point you toward gossipy locals). Something in these details will often catch the imagination of a student whose bent is linguistic, or literary, or historical, drawing him into the conversation. It will humanize the undertaking for those who fear the impersonal austerity of mathematics. Insights, points of view, ways of proceeding, will come to lean up against one another and gradually swaddle the problem in this group's unique style and flexible language, with its acronyms and abbreviations that represent and in turn promote increasing sophistication. A solitary symbol will now invoke what once a labored description could hardly encompass. This is why it is generally hard to have a new student join after the third session.

Personae will have developed too: the doubter, conjecturer, exemplifier, prover, and critic we spoke of before, and in fact the whole cast of characters which Imre Lakatos portrayed in his *Proofs and Refutations*. It is as if the landscape of mathematical invention were furrowed with specific ecological niches, which any new company, ranging over, tends to fill.

There sits the ripening problem at the center, surrounded by probes and conjectures; and this context in turn is nestled in an expansive sense of time that only the timeless air of mathematics could have lent to its making: a setting at once leisurely and intense*.

What emerges is one or two hypotheses, about some or all of the problem, that seem worthy of proof from axioms that are themselves worthy of belief.

**A Proof Takes Shape**

A different turn of mind now comes into play, in the zeal to tie what *should* to what *must* be. Making a clinching proof may have a legal sound

---

*A visitor to a class early on in the semester might come away thinking: "They'll never get anywhere at this rate!" Were that visitor a teacher, she might add: "All well and good for an after-school activity, but with a set syllabus and external exams, this would never work in schools." It could; it has. The acceleration in a course, after such thoughtful beginnings, is astonishing. People come away knowing not only more than they would ordinarily, but with a confidence based on firm knowledge that school gymnastics never give. But of course it would be ideal were today's hurdles removed and understanding math made the goal, rather than passing tests on the names of things and the application of rules.

to it (and it does demand a critical eye and a love of detail), but the same ingenuity that freed the hypothesis from its matrix of evidence operates here. From the start, in fact, momentum has slowly shifted from listening to what the data say to telling the data what they mean. Now it settles on devising, by will and imagination, artifices for bridging the gap between the axioms with their known consequences and the outstretched antecedents of the hypothesis—which will thus be transformed into a *truth,* walking hand in hand with the others.

Different personae are likely to lead this phase of the discussion: the tempered rather than the enthusiastic. The language that carries it gradually tightens too, as the premium on rigor rises. What's important, however, is to keep in check those puritanical urges, never far from the surface: the obsessive-compulsive rage to prune the last imprecision away, until the only one left for the chop is the executioner. Let the tolerance for wobble determine the coefficient of stiffness.

Bits of the proof solidify here and there, while what's between is put on investigative hold. So a network of approaches grows, branches, and thickens, and people will find the right track, though they occasionally board the wrong train.

In a class, for example, of sixth- and seventh-graders eager to find all Pythagorean triples, several cases were stumbled on, including (3, 4, 5), (5, 12, 13), and (9, 40, 41). After some rapid calculating, a boy cried out "Yes! That's it! 13, 84, 85 does work!" With a great effort, he explained his hypothesis:

"4n is the key. Take any integer n, and multiply it by 4, so 4n. Then *a* is going to be the first odd number after 4n: 4n + 1. Then for *b*, look at the first odd number after *half* of 4n, which is 2n + 1, and multiply *that* by 4n: so b = 4n(2n + 1), or $8n^2 + 4n$. And here's my proof: $a^2 + b^2$ is a perfect square: $64n^4 + 64n^3 + 32n^2 + 8n + 1$, which is $(8n^2 + 4n + 1)^2$, and that's $c^2$! That is, c = 4n(2n + 1) + 1; and there are the Pythagorean triples!"

"What about n = 0?" the girl on his left asked.

"Well, then you get $a = 1, b = 0$ and $c = 1$, so $1^2 + 0^2 = 1^2$!"

"But where's (3, 4, 5), which we know works?" asked another.

Silence—despair—until someone else came to the rescue: "Yes! If n = 1/2, you get $a = 3, b = 4$ and $c = 5$!"

Excitement everywhere. Could n be any fraction? We tried n = 1/4 and got $a = 2, b = 3/2$ and $c = 5/2$—and indeed, $2^2 + (3/2)^2 = (5/2)^2$.

We had been looking for integer triples and now were off on a branch line with fractions. Then derailment:

"I've just figured out," said a girl, "that (8, 15, 17) is a Pythagorean triple—and that won't fit this formula at all."

"No—" said a boy, who had been silently working, "but I think I've found the formula that gets it—gets *all* Pythagorean triples:

$$a = 8 + n(n + 3), b = 15 + 20n, c = 17 + 20n."$$

In an experimental mood, we first tried n = 0 and got the desired

$$(8, 15, 17).$$

n = 1 gave us (12, 35, 37), and n = 2 gave (18, 55, 57).

But n = 3? Back to fractions again! And none of the triples that had come up at the beginning could be produced by this formula. That "all" from the dawn of excitement paled to three.

Two formulae weren't worse than none; they held out hope for an overarching or underlying solution and had given us a way to play examples and formalisms off one another—which eventually led to a breakthrough (and the deep revelation that we needed to prove both a theorem and its converse: to show, that is, that whatever formula we came up with would not only yield Pythagorean triples but that any such triple would have to follow from this formula).

Nothing that happens in this phase is any more predictable than in the last, even though the attempted proofs hold the course steady. Each wreck leads to efforts at salvage by some, and calls by others to begin all over again, veering off in wholly new directions. All search for confirmations or refutations, as mood and character dictate. Initiatives overlap, repeat, repeat with a telling difference, peter out, flourish—not a line of thought but a partially ordered network spreads to catch the developing proof.

If the proof we come up with seems just to pronounce a ritual blessing—no more than a form of words to make official an insight that deserves the real credit—this phase now ends. Proof, on this level, is the "so they lived happily ever after," which lets the child snuggle confidently under the covers. In more advanced mathematics, however, one often encounters a proof that is "incontrovertible but unconvincing" (as the Russian mathematician Mark Yakovlevich Vygodskii put it seventy years ago).*
There may simply be too many steps in the argument, so that we don't grasp the sense in a single gesture. Or we may have had to take some of those steps along the way on faith: faith in our being some day able to

---

*"Two Letters from N. N. Luzin to M. Ya. Vygodskii," in *Mathematical Evolutions* (ed. Shenitzer and Stillwell, Mathematical Association of America, 2002, p. 36).

master what we missed. We may find some artifice just too contrived or, as in computer-assisted proofs, beyond any capacity of ours. We may even assent to each part, yet not accept the whole, as can happen with an inductive argument showing *that* the hypothesis is true without giving us the least inkling why. Or the result may run so counter to our intuition that we say, with Cantor: *"Je le vois mais je ne le crois pas"*—I see it but I don't believe it.

Perhaps the most disturbing effect that a proof can have is to confirm a result so monstrous that we even come to distrust some assumption we used, which until then had seemed sound (whether it be a relative youngster, like the Axiom of Choice, or as hoary with respectability as the Law of the Excluded Middle). And whatever the hypothesis, however it was proven, even when the formal proof is as close to it as a scaffolding to a façade, a serpent whisper is always at our ear: there is the proof; the proof has foundations; the foundations have mice.

## Looking Leads to Seeing

Our validated insight perches like a bird in the great tree that is mathematics. Though the roots may be gnawed, like those of Yggdrasil, the Norse world-tree, vitality moves up from the ground into the air, and we'll tend to our gardening chores later.

Students in a Math Circle class will have been so engrossed in finding and proving that they'll hardly have given a thought to the tree—which now comes into prominence. At first it is a matter of sharpening the proof: expressing what it proves with greater economy, elegance, point— but point inevitably turns into pointing. What is the context this theorem lies in; how does the theorem illuminate the theory as well as being lit up by it? For "theorem" rightly derives from the Greek word for "seeing", and after so much looking—first passive, then active—we've now made ourselves a platform to stand on and (once more passively) take in the view. Our vision's margin moves, as the formal work is adjoined to and widens our intuition.

This may happen in the last few classes of a course. More likely, these thoughts about context continue in The Math Circle's context: thoughts alone, at home, or in touch with one another. This is why we assign no homework: the freedom of math, coupled with the interest stirred by the problem and the collegial work on it, generates continuing lines of thought in the homeplay of each—and so a starburst of new directions into the unexplored. And anyway, how could *assigning* problems be consonant with our aim to let initiative flourish and to make independent colleagues, rather than disciples, of our students?

Fitting the small piece of mathematics we were working on into a picture of the whole begins with understanding it from a less parochial standpoint. "Only three people in the world understand the Theory of Relativity," they joked a century ago, "and two of them are mad." Now it is standard fare for physics students, who make the same joke about String Theory. Having to give a cogent explanation of their work to parents, visitors, and one another wonderfully concentrates a student's mind.

Soon nothing is left on hold and all's up for grabs again, as doubts beget generalizations and further conjectures open out of this understanding. That's what it means to spiral up above the problem, the level of abstraction rising (and isn't mathematics exactly the trail left by this abstract impulse?) When you finally see the solution, you move to higher order questions, and this is how awe turns into wonder. So did the shepherds leave the stage in the York Nativity Play: *exeunt omnes in mysterium*. We are always a question away from mystery.

The delight may have been impersonal, but woven into this context are images of and questions about one's own exploring mind. Looking back on the clever, facilitating fictions we came up with along the way ("Let's just say there *is* a number whose square is −1"), we think: wasn't it all our devising? Yet there the outcome stands, apart from us, its statements quantified not by "for me" and "I exist" but by "for all" and "there exists". Did we stumble on them, or they on us? Ray Berry, the great wide receiver for the old Baltimore Colts, said that luck in a game was a matter of preparation meeting opportunity. Which of these is out there, which in here? Is the distinction after all meaningful, or does the playing blur it?

Because mathematics is freedom, we play it, as music is played. This accounts for the youthful character of old mathematicians. But behind the childlike innocent eye and receptive mind, a childish secret sharer is lodged: the source of an all too frequent collapse of play into game, and of game into competition.

## Competition

Nothing, you'd think, would be more foreign to a mathematician than transforming his art into a contest. But for many students and teachers, Math Teams and Math Contests have replaced the old Math Clubs. The change in name may seem trivial, but it reflects a widespread and, we think, pernicious change in focus. Where has the leisure gone in which young mathematicians used to think through difficult thoughts? And where the fellowship of long discussions about complex ideas? Is the goal of understanding to be replaced by merely triumphing over some roomful of other people?

We've run into two sorts of argument that attempt to justify competition: a low view and a high. Surprisingly enough, the low view is more often—or at least more clamorously—expressed. Perhaps some mathematicians, trying to live down a weedy image, take a cynical stance in order to establish their Stanley Kowalski credentials. Competition, they say, flourishes naturally among adolescents who, being scrappy louts, are compelled by evolution to establish their pecking order. Math Olympiads are rites of passage: let them compete, so that the winners will go on to do successful research and chair departments and judge competitions with the benign self-confidence of adulthood, and the losers will turn to other lines of work, since they wouldn't have been happy in math anyway, as these results show. Isn't the gorilla thump of chest archetypically resonant?

In fact, some add, math is a guy thing, like war—although if some tomboy girls want to fight too, they're welcome—but let's not feminize away this last resort of the Dionysian, and turn the logically into the politically correct. You mustn't hear undertones in all this of the Will to Power, nor see the spectacle as if filmed in Nuremberg. No one is being shot. There is no backstreet mayhem while the banners brightly fly, and we have the assurance of the winners that no one minds losing.

The arguments for competition based on the nature of mathematics are even sadder: math, after all, isn't very attractive—is repulsive, in fact—to most people. People do, however, like sporting events and contests. "The great gift of competition is that it develops excitement," said a proud impresario of these events, as if mathematics had no intrinsic excitement itself.

Isn't this a little odd? If someone really thinks his profession not worth pursuing, why proselytize—for the company that misery loves? Actually, most of the people who make this argument are secretly infatuated with mathematics but have given up trying to persuade others of its appeal. Playing math down in this way could also be seen as a form of modesty: "The proper behavior of a !Kung hunter who has made a big kill is to speak of it in passing and in a deprecating manner."

And, of course, some people actually do hold a low view of mathematics. For them there is fortunately no tension with the idea of contests. These are the people who see mathematics as a required subject, a thicket to hack through before being allowed to get on with one's life. We've come across this view often in our teacher training. Riding up in the elevator once to the first meeting of a course we were giving to high school teachers on The Math Circle approach, we heard one say to another, "I didn't know you'd switched to teaching math." "Oh yes," she answered, "I hate it, naturally, but the pay's much better." People who think like this know that winners of any kind of contest automatically step into the elevator to success. If the subject happens to be math, they now have in their hands a certificate that important people who haven't the least inkling of what math is about will nevertheless—*or therefore*—value: *Time* magazine described the list of winners of the Putnam Competition as "a social register of the aristocracy of arithmetic." Mathematics is a way up from the depths for anyone having this quirky talent, whether immigrant's child or janitor at MIT.

There are nobler reasons for valuing competition. Consider first the view that mathematics will timelessly triumph, rising like Mont Saint-Michel from anonymous lives and the accidents of place. We've heard before the ascetic view that it is the great insights that, justly, become part of the eternal structure, not those who had them. Such insights are usually called forth by deep or unsolved problems, neither of which will appear in a contest's format but, under the pressure of time and competition, a new and elegant solution could emerge and become immortalized. It has even been claimed that contestants will acquire permanent habits of perfection: "[The Putnam Competition] consists of problems for which an elegant solution is guaranteed to exist," says Bjorn Poonen, four times a Putnam Fellow. "People who spend a lot of time working on such problems become conditioned to seek and even expect such elegant solutions. . . . They tend to seek not just any way of explaining something but the best way."

Still, the spirit of competition can be prolonged past contests per se by offering large cash awards for solving open problems—and the reward of prestige is always there. Why take this away, when it not only appeals to human nature (a rather low view) but may indeed yield solutions that might otherwise not have been born? So bizarre is the relation of the synthetic to the a priori. The solutions, not the grubby inducements to them, are what matter; art satisfies art, whatever else may satisfy the artist. Needless to say, this argument ignores the complex relations between art and artist, and takes as best how things merely happen to be.

The claim that contests catch the attention of those who might not otherwise have taken to mathematics, sounds like the low claim that math is too cold in itself to excite anyone—but really follows a different line of thought. If we teachers haven't found how to make a piece of mathematics attractive in itself, why not turn it into a puzzle, whose playful challenge will do the attracting? No one feels threatened by a puzzle, and a good riddle can even engross attention in a manner resembling the mathematical madness. Those who organize meets with intentions like these do so in the well-meant but mistaken belief that grafting the playful element in human nature onto the sturdy root stock of aggression will produce powerful creativity. But by the easiest of linkings-on, puzzles become games, games contests, and contests competitions—and the fun somehow leaks out along the way. How sensible to shorten the endless afternoons of sandlot baseball to a sparkling and official Little

League—which turns, before you know it, into the pro game, where the baseballs never run out but the contracts do.

Yet don't these competitions make all of their participants better? Training for them in the public gyms of math camps, or on your desktop mathematical Bowflex, can't but increase your mental athleticism: skills get a workout, stamina increases, the threshold of frustration steadily rises. A carefully calibrated diet of problems will teach you to think strategically, and at the same time fill up your handy bag of tactical tricks. A problem, however, with one-size-fits-all gimmicks, like "add zero creatively," is that they smooth over suggestive peculiarities, making all problems seem alike—and at the same time hide the deeper structure from which these hints come: in this case, the key role that scaffolding fictions play in mathematical creation.

On a higher level, we've already had testimony that contests sharpen the eye for elegance. More: "They give a glimpse of the beauty of mathematics in general," said another Putnam winner, Ravi Vakil. How does such an exam frame this glimpse? Moments of sublimity teeter on perilous edges; Vakil went on to say of the exam that "awe, respect and terror all apply." Terror of the great heights and chasms that the questions opened up? Or the much more familiar terror of being weighed in the balance and found wanting? Is it this we want our students to prepare for? And if a notable winner felt terror . . .

Jeff Lagarias, another winner, said that from the age of fifteen he had hoped to be a mathematician, and success in the Putnam gave his confidence a vital boost. But what did it do for the confidence of the many who didn't win? We've been assured by experts that adolescents, whose self-esteem we had pictured as fragile, are actually resilient by nature. Being experts, they must be right. We wonder, though, where they found their evidence. Perhaps in the well-known bonhomie of the cohort in math training camps—but people in company generally tend to talk up themselves and one another. We don't see them alone in their rooms, nor overhear their thoughts.

After all the window-dressing arguments are gone through, the real agenda of those who promote math competitions stands clear: to cull out the best from the rest. It isn't just *Time* magazine that uses terms like "an aristocracy of arithmetic": "Each summer six math whizzes selected from nearly a half-million American teens compete against the world's best problem solvers at the International Math Olympiad" (from the publisher's description of Steve Olson's *Count Down: Six Kids Vie for Glory at the World's Toughest Math Competition*). "The top 150 AIME participants participate in this elite exam" (from Paul Zeitz, *The Art and Craft of Problem Solving*). The analogy to chess rankings is never far away: masters, grandmasters; local, regional, and national champions—and a world champion at their apex: all determined by competition. Were math as combinatorial as chess, then some day we should have a Deepest Blue churning out proofs. At least it would blow all these head-butting matches away.

The impetus behind competitions isn't the kids at all but adults with a program (such as those who approached us once with an offer of serious funding, on the condition that we would henceforth be "top down"— which meant, it turned out, that we were to dump our random collection of students and take only the top few who *deserved* to encounter real math). In the days of the British Raj, proficiency in the classics served to rank future colonial administrators (as we mentioned in chapter 1, it showed you could put up with anything). Now, in a different kind of empire, proficiency in math is seen as displaying qualities of intelligence, diligence, and character. Were very few places available for vast numbers of college and university applicants, a Mandarin sorting exam might be appropriate; but this just isn't how things are. Perhaps these competitions are a long shadow of the Soviet Union's vast populations and limited resources.

But let's for a moment suppose, with these contest organizers, that there *is* such a thing as mathematical talent; that we needed for some reason to select for and display it; and that acronymic competitions (AHSME, AIME, USAMO, ARML, IMO) were the way to do this. How, in the name of reason, could we then hold training sessions for the candidates, and publish books of tricks and tropes, and compilations of past problems? Since the aim of the exam was to reveal *talent,* this training and coaching could only blur the results, and let hard plodders and shallow quick studies rob the true geniuses of their rightful rewards.

Possibly the coaching enthusiasts hadn't thought this through. More likely, a very different intention underlies their enormous effort. For what you hear from them, again and again, is their list of winners: my team beat theirs. The voices seem to come from the locker room or the frat house. "We can't have three years in a run as good as Team Canada," the leader of the American contestants at a recent International Mathematical Olympiad said. Why not? Is Canada now an enemy? Is a conference title at stake? "You know, the University of Nebraska fired their football coach after he went 10 and 3," he pointed out. We can't have that bad a record here: reputations are on the line. Status, status, status

This may just be the excitable rhetoric of coaching, and behind it a familiar and wholly understandable devotion to one's little community. These students are in our care, and by sheer good luck they happen to have gigantic potential (this new weekly group, this team, this year's summer camp), so that it is only right to develop it. Besides, since those other teams are being coached, our real mathematicians wouldn't show to advantage on a common test, despite their aptitude, were they not coached too. And what if some of ours do well, who by innate right shouldn't—don't they deserve to profit from their effort? Anyway, we may have misread their talent—and there's always her home situation, and his health, to take into account. So compromise and confusion replace clarity of purpose, and when this or that kid wins, we bask in the reflected glory and never lack for retrospective evidence that they were so destined (the best all along, or the romantic exception).

Beyond the glory-basking, the unexamined pleasure of sharing in a passing cohort, and the satisfaction of having selflessly served the higher good of mathematics, subtler factors may contribute to promoting competitions. At one extreme would be a sense of covert power, in being mentor or judge; at the other, a chance for the socially inept to be for once the life of the party, and for the dedicated to experience the distanced intensity of intellectual work, within the constraints of coaching and competing. One may not even be aware of such relations, save in vivid memories of sessions with a particular student, long ago, on a particular theorem. Between these extremes, a wide variety of outsiders will have found a place where they can be—by right. The more competitive will enjoy relegating others to somewhere outside their snug elitism. Once a tradition of regular meets has been established, how hard it would be to give up the camaraderie that goes with them! The regiment may be redundant, but what pleasure swapping stories in the regimental mess.

Put that way, it all sounds harmless—irrelevant, unnecessary, but unable to damage mathematics. But just as for many people baseball means the World Series, for many students math means the Olympiads. There is as much of a disconnect between the culture of math competitions and the civilization of mathematics, as between the school perception of what math is about, and the work of mathematicians. The contest mentality seems founded on bypassed assumptions and a studious turning away from mathematical practice. Faced in this different direction, it cultivates a charmless style of human exchange.

The first of these bypassed assumptions is the mistaken view that what best serves mathematics is a narrow cast of net: contests that call to a specialized sect, from which steadily fewer are chosen. But wouldn't it be better to let as *many* people as possible, brimming with self-confidence, probe these mysteries? We aren't Macbeth's witches who could look into the seeds of time and say which grains would grow and which would not. How tell if this eight-, this twelve-, this twenty-year-old will continue as she is now? Surely the results of answering five questions in forty-five minutes, or even six questions in five hours, make a paltry oracle. Think what would have happened had our native language been learned in this style, teams coached in problem-solving converging on a decisive exam? Each state might then have its linguistic hero, while the rest of us lived functionally mute—as do so many in math. This view of contests takes competition as natural to humans, and evolution as tuned by survival of the fittest. Hasn't evolutionary theory grown subtler, with a niche-filling plurality displacing the single supremacy model?

A second outmoded assumption is that young people take losing in stride. Consider the "stereotype threat" described by Steele, Aronson, and Spencer. When a math test was given to a group of women who were told, ahead of time, that men and women did equally well on this test, they did just that. When told it was a "math ability" test, they assumed their ability going to be measured against men's, and flopped. When white male math students were told that Asian-Americans "did really well on this," they too nose-dived. *Amour propre* is as sensitive as a mercury switch: a little jiggle can turn off your attachment forever. If you say that contests, and the run-up to them, are meant to toughen you up, it isn't clear to us that thickening the ego is the surest way to let imagination and conjecture loose. Sensitivity and vulnerability are two sides of the same coin: may the judgment of those who lead the young be as good as their will.

We said that the contest mentality turns away from mathematical practice. Since the days of Laplace, readers have been asked to build an argument's connective tissue: the more they do this the stronger and suppler will be their command of the ideas. But once again means inexorably become ends—let's call this "regression to the means"—and people delight in posing problems for their own sake. Like crossword puzzles, these can be an innocent pastime for those who want time to pass, but problem-solving can diffuse concentration and lose you a sense of the whole. Contests also turn away from the doing of mathematics by shortening the arc between problem and solution. For more than a century, now, challenging problems done under time pressure have been seen as a way not only to skim off the cream of the latest crop but to re-create in miniature the essence of the mathematical enterprise. It isn't rare, however, for someone who thrives under these circumstances to be at a loss when faced with the longer trajectory of a research problem. So batteries recharged too frequently will eventually fail to hold their charge at all.

In *The Mathematical Experience*, Davis and Hersh point out that students have a very strong desire for immediate comprehension, which may ultimately be debilitating: "If I don't get it right away, then I never will, and I say to hell with it." This is a weakness that contests cater to and reinforce. Going more deeply into the problem, William Thurston, in his essay "Mathematical Education", reminds us that many top mathematicians were not good at contests, which put a premium on speed at the expense of depth, and that

William Thurston's
" Mathematical Education"

> These contests are a bit like spelling bees. There is some connection between good spelling and good writing, but the winner of the state spelling bee does not necessarily have the talent to become a good writer, and some fine writers are not good spellers. If there was a popular confusion between good spelling and good writing, many potential writers would be unnecessarily discouraged.

How does the culture of contest promote a charmless style of human exchange? Instead of appreciating, encouraging, and even contributing to the work of others (so much the norm in math department common rooms that it has come to be reflected in their open-plan architecture), the participants in training sessions tend to vaunt their prowess and guard their gains—as if insights (so many of which come from stereoptical looking) were merely property. Having long since stopped beating students to make them learn, why do we still encourage them to beat one another?

We've spoken before of the passive and active sides in the mathematical makeup: quietly taking in the lay of the land, actively imposing your will on what you find there. It is as if in the world of competitions, the aggressive component had broken loose and now ravened about on its own, with math as no more than its excuse. You naturally want to sharpen your mind, and now find it socially acceptable to use your companions as grindstones—but a variety of quieter temperaments are put off by such an atmosphere of estrangement, and everyone must find it more difficult to take those risks from which the profoundest revelations so often come.

Hugo Rossi, of the Mathematical Sciences Research Institute, rightly says that "competition is an essential motivator for some people, and irrelevant or even detrimental to others." Those who profit from it (keep in mind the possible male-female distinction we spoke of before) should have every opportunity to indulge their fancy: this is part of the productive variety fostered by the freedom of mathematics.* Our objections are to the decay, within the competitive camp, from the pursuit of math to the pursuit of prizes (another instance of regression to the means), but even more, to the current tendency to take this inclination of some as the standard for all. We don't pretend that because this mathematician is a vegetarian, or that one a player of the shakuhachi, all should be.

---

*If there are to be competitions, might they not be in a more fruitful style? Barry Mazur asks why all—or the most telling—questions couldn't be of the form: "What do you think about this?" He calls the present form of competitions "closed": here is the problem we have made for you, solve it. He suggests instead an "open form": tell us something of your own; surprise us. This would put competitions more in line with those of ancient Greece, where playwrights submitted their works for the prize of being performed on a great holiday.

Because that great algebraist Cardano slashed the face of a man who cheated him at cards, doesn't require every probabilist to carry a knife.

The underlying mistake, it seems to us, is thinking that math is the measure of man. On the contrary, man is the measure of math, and each of us measures it differently. From these differences, the architecture of understanding grows. And if you are one of those for whom competition is a vital spur, why trivialize your urge by competing merely against others, when there is math itself, with its endless open questions, to take on? Isn't man against the gods a higher drama than man against man?

Why should we hand over the way we learn mathematics to the less attractive parts of our nature (greed for conquest, a shallow view of self and of mathematics, and a dislocated sense of the good)? Why not let math evolve with us, gaining in imagination and scope, and ever more buoyant in the freedom which is its essence? Let's treat it seriously, and the people doing it as equals. Let's play with their strengths, rather than on their weaknesses. Let's invent our way together toward discovery.

# nine

# Filling in the Details

## Where's the Kit?

If you've been waiting impatiently for a neat package of curricula, in-class exercises, and homework problems, all ready to photocopy, with instructions on what to do in all the contingencies that come up when working with others, you're about to be disappointed. Mathematics is freedom, and part of the exhilaration of doing it is that you just don't know what will happen next. You've set up a destination in your own mind, you have a general map of some routes to it, but between setting out and getting there are all the adventures that mind and the world make for each other—not to mention the differently angled wishes and wills of your fellow voyagers. You may lose the way or find a new one; you may end up somewhere else; you may stumble and slide or run effortlessly along the shortest path. What we have for you is a representative skeletal outline of one course and, from the journal of some others, notes that tell you more about the actual doing than an outline could: assorted vignettes, anecdotes, reflections, and bits of advice. In prospect, the engineering of any great undertaking can't help but be piecemeal: that's what it means to be synthetic. It is only in retrospect that the falls are seen as fall-lines, and the architecture as a priori.

> Pure mathematics is the magician's real wand.
> —Novalis

## Courses

Many different courses have appeared in The Math Circle since it began in 1994. Some topics suited all ages, when adjusted to the right level of sophistication. These included:

Cantorian set theory

Group theory

Knots

Map coloring

Number theory (always focusing on a specific issue, from congruences and continued fractions, to which numbers are the sums of two squares, and quadratic reciprocity)

Probability

Sequences and series

Straight edge and compass constructions

The Pythagorean Theorem

Tiling

### *Courses for five- to eight-year-olds*

Are there numbers between numbers?

Arithmetic machines

Arithmetic on different bases

Clock arithmetic

Eulerian and Hamiltonian circuits

Figurate numbers

Fractions and decimals

Functions and their graphs

Interesting numbers

Kinds of numbers

Prime numbers

The Euclidean Algorithm

The Mathematics of Origami

What is area?

### *Courses for nine- to thirteen-year-olds*

Angle trisection—or not

Calculus via maxima, minima, optima

Complex numbers

Construction of polyhedra, and the Euler Characteristic

Convex figures

Equidecomposability

Game theory

Graph theory

Information theory

Interesting points in triangles

Logarithms

Mathematical games

Mathematical modeling

Numbers and nimbers   nimb-ers?

Pascal's Triangle and fractals

Projective geometry
Propositional calculus
Random walks
Solution of polynomials by radicals
Styles of proving
The Divine Proportion
The Fibonacci sequence
The Intermediate Value Theorem
Visual proofs
What Is i? *Yeah? What is it?*

### *Courses for fourteen- to eighteen-year-olds and up*

Algebraic geometry
Classification of surfaces
Complex analysis
Computational complexity theory
Fractals
Group theory, topology, and physics
Hyperbolic geometry
Mathematical logic and Gödel's Theorem
Measure theory
Proofs from The Book
Quantum mechanics

## Sample Outline of a Middle Course:
## Interesting Points in Triangles

### *Some Aims*

1. To explore and develop two-dimensional spatial intuition
2. To encounter and invent deductive proofs and the idea of an axiom system
3. To develop the play between insight and proof
4. To unveil the endless layers of structure in any topic

## Format

Here is a triangle ABC (best drawn large, and as obtuse): what you see is what you get.

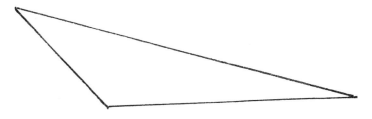

A casual, seemingly unrelated question: are there any points exactly midway between the two fixed points, P and Q?

If the midpoint is offered: is it unique? Lead them to the perpendicular bisector of line-segment PQ (if offered as a line-segment, lead to a line; if anyone points out that there is a whole plane of bisectors, note it and return to the plane of the triangle).

So we have a perpendicular bisector m of AB. And of AC? Yes, call it n. Do they intersect?

We've reached a bristle-point—the discussion can, fruitfully, go off in many possible directions: if (as we mentioned earlier) you draw m and n as parallel or asymptotic to one another, intuition will be outraged, and interesting discussions—which may (fruitfully) displace the initial aims—will begin, such as: how trust ink or chalk representations? Aren't we talking about ideal lines and points, which our drawings only suggest? How, why, trust the visual at all? Relation of math to the world, or: launching into a proof that these two perpendicular bisectors must meet.

This is an important and ambitious project, which will give the students their first taste of some Euclidean axioms and of deductive proof (with issues here too of elegance and economy, and of intuition leading the way). Let axioms arise on a "need to have" basis: only those definitions and axioms should surface that are needed in order to prove intersection of these two lines (so the concept of "parallel", for example, will come up). Always let the students struggle to find suitable definitions and axioms; take the latter as desperate measures, to stop endless regression. Note: take SAS as an axiom, and when ASA is needed, take it as an axiom too (saying that we might have been able to deduce it from what we have), unless it seems right to digress into the proof of it. It is very rewarding to have confirmed, after much work, that m and n indeed intersect—and begins to adjust the relation between intuition and proof.

With m ∩ n = O, ask about r, the perpendicular bisector of BC. Draw it as intersecting m and n in different points P and Q, pause, and let conjecture and intuition have their say.

Their discovery of the circumcenter via transitivity alone (since O is on m, it is equidistant from A and B; since it also lies on n, from A and C—hence by transitivity, it is equidistant from B and C) is elating—it may also misleadingly suggest that all proofs will be this short or simple.

Having started with an obtuse triangle leaves O vividly outside it. Where will it be in an acute, in a right, triangle? Bristle-point: shall we look at its trajectory, or think about continuity?

Are there other such points of concurrence? Are there other interesting triples of lines to concur? Angle bisectors tend to come up next, and it is pleasant if they do: for their concurrency is moderately intuitive, and a proof advances the grasp of Euclidean geometry, but not uncomfortably.

Medians are unlikely to be thought of: it would be well to suggest them here. Bristle-point: proofs of their concurrency abound, with different flavors (some leaning on physics and centroids, some descending from generalizations such as Ceva's; some are ponderous, as in Euclid, some daring and dapper, as from projective geometry—the rest of the course, or a course in itself, could be devoted to this variety and its implications).

Now altitudes and the orthocenter. This seems the least intuitive concurrency, and reaching for proofs the most strenuous. It comes as a salutary shock—the shock of elegance—as we mentioned in chapter 8, to prove their concurrency via seeing them as the perpendicular bisectors of the triangle similar to, twice the size of and enclosing (inverted) the given triangle.

Bristle-point: what about concurrences of analogous lines or planes in tetrahedra? } yes!

The second stage of the course: we now have circumcenter O, incenter P, centroid G, and orthocenter H. Is there a structure hidden here? Yes: the Euler Line, on which lie O, G, and H. While this collinearity may be seen, its proof is unlikely to be devised. Following Euler's, however—with as many of the steps as possible found, or justified, by the students—will raise their level of mathematical sophistication.

Third stage: more, and deeper, hidden structure: the nine-point or Feuerbach Circle (whose center lies on the Euler Line!). Again, they will more likely be assisting at than inventing a proof, so encourage doubts, criticism, variations, that their savoring may be that of apprentice rather than spectator. Triangle points

More and more points show up on the nine-point circle (such as points of tangency with each of $\triangle ABC$'s three exocircles, and with its incircle—and in fact an infinity of others); more and more invariant points associated with a triangle may now arise (the Symmedian Point, Brocard Points . . . ), and other lines (the Soddy and Gergonne Lines), and other triangles: Brocard Triangles, the Euler-Soddy-Gergonne Right Triangle. Endless directions in which to go from here: into higher dimensions or on to n-gons, and the significance of collinearity and concurrency could equally well lead to projective geometry and duality. What began as no more than a triangle ends with invisible structure becoming palpable to the eye—but more significantly, to the mind.

Note: knowing how rich the harvest can be, it is important to restrain oneself and let the pace of the course be andante: better that the students end up having discovered for themselves no more than the circumcenter, than to leave them drowsy with other men's flowers.

## Piecemeal Advice

Here are some notes we've handed out to new colleagues, a week or so before they begin to lead a class.

## Perilous Turnings and Pivotal Moments

***The Class***

1. "I heard myself saying 'Wait a minute, let me show you how to do this.'" The moment you become aware of its effects, you'll have broken the habit we have all developed of turning our students into spectators. The most fundamental message of The Math Circle is that you learn math by inventing or discovering it yourself.

2. A frustrated student gives up. "I'll bet *you* know the answer!" You probably do, or know pretty well how to get it, but that's of no importance. The fiction that you were all working on this together—that the problem, not accidents of knowledge, mattered—has been broken and needs to be restored. A good reply is the honest one: that the answer is still not known to the group, and that they'd no more want to be told it than a sprinter would like someone else to do his training or run his race for him; that does nothing at all for his muscles or his self-esteem. The Math Circle's motto comes from the eighteenth-century physicist, G. C. Lichtenberg: "What you have been obliged to discover by yourself leaves a path in your mind which you can use again when the need arises." It doesn't matter who—or indeed if anyone—knows how to do this: it's our turn now. It is only the temporary fatigue brought on by frustration that makes you want to be told the story instead of writing it yourself. Regroup, refocus, return to the details of the problem at hand and ask: What exactly stands in our way?

3. Up the airy mountain. The conversation has become more and more general, the fecund details of the problem blurred away. This is inevitable, given the abstract impulse native to math, and can lead wonderfully on—at the right time. Make a public record (hand-outs, blackboard) of these paths leading away,

so that the air is cleared, no one feels cheated of his insight nor distracted by the other vistas opened up, and then come back to the problem at hand, your imaginations refreshed by these opened windows.

4. Down the rabbit hole. The conversation has turned in ever tighter spirals into details of details, until all sight of what you've been doing it for is lost. Judge what the right moment is to surface again, survey how far you've come and where you're going. A systematic sketch of gains and losses, asking now for strategy and tactics, renews the enterprise—but don't make it too tidy, lest a mechanical ideal displace the imaginative.

5. Careening away in the wrong direction. Someone comes up excitedly with a conjecture, which is just dead wrong. Fine: let it play out. "That's interesting; what does it imply?" Or with equal excitement: "Let's try it on an example," or "Fascinating! What makes you think so?" Many a time real insights come from errors; in any case, it sharpens the critical faculties of all to probe at a conjecture, right or wrong. Keep the risky character of this enterprise alive; it makes the whole undertaking more exciting, and when a sound path is eventually hit on, the experience is all the more stunning. Out on the frontier one tends to make wrong guesses more often than right ones. If one wrong turn follows another until the initial issue is lost in the fog, nudge the conversation back in a fruitful direction— but nudge as unobtrusively as you can (coming up with a suggestive example for them to work on is always helpful).

6. "What's the point of doing this anyway?" The course or the class began with an abstract question (Can you tile a rectangle with incongruent squares? What is $i^i$?) and everyone plunged enthusiastically in, caught up in the spirit of the hunt. But once bogged down, with self-confidence on the ebb, it is natural to turn on the source of annoyance and try to dismiss it. Suit what you say to your temper and theirs, but points worth making are that anything leads to everything; that math's history is one of research as an end in itself always turning out later to bear significantly on understanding and controlling the world around us; that as an art the point is in the beauty of the revelations; that it is, after all, frustration which has undermined our efforts and made us question their value. Time for a pause and a step back in order to leap forward.

7. "What time is it?" This question stands for disengagement, which can become contagious. If the students are very young, the moment has come to drop whatever you're doing (no matter

how interesting it has become to you) and play a game (function machines) or tell jokes or anecdotes, that will have some later value. If the question comes from a student in an older class, feel free to say that there's no compulsion to stay: The Math Circle is supposed to be fun, and if you're not finding it so, we'll think none the worse of you for leaving. But the implication of the question must be addressed on the spot. And check your watch: class may have ended ten minutes ago, and your questioner has just noticed the crowd at the door.

8. "*I* know, *I* know!" Eagerness and lack of self-awareness may excite people of perfectly good will to want to answer every question, even the rhetorical ones. "I'm eager to hear your idea, but it's Alice's turn now." Keep the shares in the conversation evenly divided; arrange auctioneer's signals with the shy. Even if an enthusiast bursts into tears (as five-year-olds have been known to do) at not being called on, go on blithely to others and return to him, when he seems under better control, with a new question.

9. A student runs up, takes the chalk from your hand and begins to write out a conjecture, an approach, a proof, a diagram, a counterexample, on the board. This is what you've waited for, when they begin to take over. It won't be smooth sailing, but better a rough voyage with them at the helm than a smooth one steered by you, which they sleep through in their cabins. Encourage others to come up to the board too (standing helpfully near the shy and away from the bold); if it is always the same student who rushes up, explain in an age-appropriate way about taking turns.

10. A voice from the back of the room. If you decide that well-behaved parents may audit a class with their children, tell them ahead of time that auditing means just that. You may nevertheless find a parent, or a visitor, breaking in to a discussion with "That isn't true!" Parents are less good than their children at understanding your role as Sherpa or dumb friend. "I have a proof" is another familiar baritone aria. Smile and congratulate the latter, smile and raise an eyebrow to the former, tell both after class that the students have right of way.

11. Children as hand puppets. "My daughter wanted to say . . ." "My son really means . . ." Well-intentioned parents sitting beside their offspring can not only stifle their initiative and confuse them about what they thought, but rock everyone back on his heels. A good policy is to ask parents, who stay after all

*What a phrase!*

with their offspring's permission, to sit at the back. If they natter away noisily with one another, don't hesitate (looking at the ceiling, perhaps) to ask them for the same attention their young are giving. If a child's shyness really makes it seem right for the parent to be alongside, ask at the first of these hand-puppet moments for the child's opinion. Some of the less perceptive parents need to have the idea spelled out for them after class. Some really want success and recognition more than learning— that's worth a telephone call before the child is pushed into deception or despair.

12. The breakthrough. Suddenly the light dawns—for all, or some, or someone. Celebration is in order, but not excessive praise. The focus has to stay on the issue at hand, the insight pinned down, explored, made clear by its discoverers to all, its consequences turned to. Too much praise not only loses sight of the math but is condescending, and it revives the competitive spirit we so much want to divert into the energy of inquiry. It is inevitable, too, that praise of one is heard as diminishment of another. Mark the moment and move on: for small victories, "Good. Now—." For large ones: "We've done it! Now—" and on to consequences or further conjectures.

## The Teacher

13. The Pied Piper. You want to be followed—but only so far. Your energy, enthusiasm, optimism, and good will are crucial at the beginning; they give a class its buoyancy and make everyone rightly feel part of an important enterprise. Gradually, however, you want them to take over: you become their secretary at the blackboard, or referee, or companion climber, or their slow friend for whom everything must be spelled out. Your radiance remains, but in the background. You want them to leave correctly thinking: "We"—not "I" or "He"—"figured it out!"

14. No idea. A student comes up with a conjecture or a suggestion or a line of thought that leaves you simply without a clue. Follow it out with the class; feel not the least twinge of embarrassment in admitting you don't know whether this is a sound idea or not; say you'll work on it and hope they will too. If you get nowhere with it at home, try a friend, a colleague, or the Internet. At the next class, report on your efforts and find out about theirs (you're all now fully engaged in the real work of doing math). If they came up with nothing, ask a question that will set them on the new track you've found.

### The Students

15. "Well that's because I'm a genius and you're not!" Especially in new groups there are likely to be those who have to prove their credentials at the expense of others—and nothing can kill a class deader. You have to address put-downs as soon as you hear the least whisper of them. An upbeat, serious remark to the class as a whole sometimes works best: "We're working together here, in good spirit, and math is too profound for our little egos to get in the way." A word afterwards with the self-aggrandizers may be in order; it isn't rare for them to change when they realize that this isn't the culture of the present society. If nothing works, it is vital to ask these disturbers of congenial conversation to leave. If you charge for attendance, refund their money in full.

16. Clever Hans. Children are masters at reading subliminal signals: your hand moving toward the board as their ideas approach what you have in mind; the half smile, the half frown, that betrays what you think of their efforts. Most teaching is acting. Try for uniform enthusiasm or neutrality (as suits you), remembering always that you want them to do the discovering—and of math, not of your secret knowledge or opinion. If you're going to play favorites, play it with everyone.

17. "Do this thing to that thing . . ." Lack of experience, lack of vocabulary, sheer excitement, will make students blur their insights and explanations. Judge when to ask for more rigor: not so soon as to kill initiative nor so late that all of you have become hopelessly confused. "So this thing . . ." "You mean this median?" "Yes, well . . ." "OK, give it a name." "This median m . . ."—and you're on your way.

### The Material

18. The shuttle. Not just your students but all mathematicians shuttle endlessly between intuition and rigor. You need to decide when one becomes excessive and lead the conversation toward the other. "All right, I get it" is a good signal that an approach is rigorous *enough* (you don't want to legalize it to a standstill, and can always come back to snap up unconsidered trifles). Saying "I'm confused—this is getting too blurry for me" works well when intuition runs headlong away.

19. Was the class, was the course, a success? Often the best guide to that is whether you enjoyed it. But cast a mental eye back over what happened to see if anyone had been left out of the fun.

Did you get to where you had wanted to go—or to somewhere else equally valuable? Above all, did the students singly and together catch a sight of the matchless beauty of math?

## From a Journal

The outline of "Interesting Points in Triangles" tells you what you need to know about the mechanics of a course, and the notes for new teachers above add substance. But they don't touch on what actually happens, and how the mechanical becomes organic in the presence of personalities. Lest you think that our classes are symposia of Platonic beings rather than quirky people, here are extracts from a journal tracking the difficulties that arose in three different sections of a fall course on polygon construction. Sections A and B were ten-year-olds, section C was eleven to twelve.

## SECTION A

### Class 1

A class of ten, including five pals from the same school, all clearly used to having their own way. They ignored the other kids rather pointedly. So now we have a social issue to deal with as well as a mathematical. Handed out compasses (note: they were cheap ones, and that's a false economy—too stiff or too flexible to be useful, with a clip which allowed the pencil to slat about), and gave them a whole hour to fool around with making patterns. Neat fingered girls, C and S, produced Gothic arches, quilt patterns, yin-yang designs. J decided to do them freehand, because his technique (hold the compass still, rotate the paper) wasn't too effective. Boys with poor small muscle control gave up quickly—perhaps they will be the force that moves us to abstraction. N gave up, H and Z went to the board and used string and chalk. Question: is it worth having a practical component? Yes. The Platonic idea is, of course, in the head, but it's important that the hand get a glimpse of what making actually is.

### Class 2

Is it spoiledness, TV jump cuts, overscheduled lives or what, that makes modern youth so ready to give up? J decided it wasn't worth trying to bisect a line segment before the problem had even been posed. I raised the question of the sum of a regular polygon's interior angles—triangle, square, then went on to hexagon, because I thought it easier to calculate in the remaining time. Class instantly leapt into pattern-searching: $3 \times 60°$, $4 \times 90° => 6 \times 120°$ "because they go up by 30"—but then somebody noticed the 5 was missing, and a shouting war broke out between adherents of 120° and 150°. N cut a hexagon into 6 equilateral triangles, so won the war for 120°, but their question is always "what's the right answer," not "how do you know that's the right answer"—and "it *looks* equilateral/parallel/equal" is totally satisfactory for them.

### Class 3

The Gang of Five was overexcited. Sugar may not be a cause, but bringing candy to class and not sharing it equally certainly is. I ban eating in class with a vivid fable of cockroaches among the overhead steam pipes. A's mother, while people are settling in, fears he has not told me about the multiplication pattern he has discovered: $6 \times 6 = 36$; subtract 1 and $35 = 7 \times 5$; subtract 3 and get $32 = 8 \times 4$, subtract 5 and get $27 = 9 \times 3$. She wants admiration for her boy—he looks embarrassed, but clearly wants admiration too (especially because the Gang pretends he isn't there). Question: stop the topic we're working on and let A show this for social good (but will it do the trick)? Get the whole class working on trying to figure out how it works (if x is a square number, call it n—then $(x+1)(x-1) = x^2 - 1 = n - 1$; $(x+2)(x-2) = x^2 - 4 = (x^2 - 1) - 3 = (n - 1) - 3 \ldots$)? While distracted by simultaneously hearing this saga and trying to decide what to do, the class has arrived and gotten out of hand. Z announces that he remembers nothing of the previous classes, no one has brought in a compass (my solution to their complaints about the crummy ones I'd provided was suggesting they bring their own). Figure I'd better save A's hypothesis for later, and get the show on the road. Dropped my voice a fifth and said that if we were going to get to the polygon stuff we'd have to use our time better—and it *worked*. There was real concentration on how to create a square. C got a right angle by bisecting a line segment. A had an interesting idea of drawing a circle and constructing tangents to box it in—except he couldn't find a unique point. Then N made a right angle, swung an arc from the angle to hit the two sides to get adjacent equal sides, and then found the fourth vertex by swinging arcs of that length from his two new vertices.

A day saved from chaos—until the end of class when N had to throw his shoe to knock down the paper airplane that had got caught among the steam pipes and it almost landed on A. . . .

### Class 4

Three new kids arrived in the class, so a chance for people to demonstrate to them all the things we could construct. A new octagon construction suggested—"make a cross of five squares and connect the points"—looked very good, but J pointed out that the hypotenuse of a

The pattern I've played with {

right triangle has to be longer than its sides, so it wasn't a *regular* octagon. A great event —a revelation to the "it looks like it so it is" brigade, who for the rest of the class demanded proof; but usually by asking J or me to provide it. J is attentive, dogged, and imaginative—a real spark to the class and a great help in showing that the topic is not the sole property of the teacher.

Good discussion arose about pentagons—first appearance of conditional thinking: *if* we had a 36°–72°–72° triangle, giving us a length of x, would that mean that having x we could create a 36°–72°–72° triangle?

### Class 5

Lots of schools apparently use April Fool's Day as a Saturnalia, so there was a constant undertone of giggle, and we never had more than half the class involved in any discussion—so as people rotated into consciousness there had to be a lot of recapitulation, which bored a previous conversationalist into dropping out. One scary moment: A was very disturbed by our finding out that $\sqrt{5}$ wasn't rational. He kept saying, "Tell me what it *is*!" and got red in the face and close to tears at its refusal to turn into a fraction.   A true Pythagorean

### Class 6

"This is the best class we've ever had," say N and H. Why? We were actually making a pentagon—which they found *very* hard, but they knew what they wanted to do, stuck to it, and ended up with perfect, if grimy, pentagons. We've really established why $\frac{\sqrt{5}-1}{2}$ is crucial, where it comes from, how to construct it—and they were confident that if they just tried they would get a pentagon. J got a nonagon—twice! Other factors: Z and M weren't there, and since they are often not there, they usually haven't a clue and so distract their neighbors.

### SECTION B

### Class 1

I started by drawing a really crude triangle on the board. "What can you say about this?" Chorus: "It's a triangle." (In tone of *duh*—how old do you think we *are*?) G: "It's a pretty pathetic triangle." So we set up a spectrum from *pathetic* to *perfect* (Platonic) and *constructed*. Lots of good comments—that it was impossible to write "perfect" perfectly on the board; that one must have the Platonic triangle in one's mind to know that the pathetic triangle was pathetic. Oddly, this group is much more manually dexterous with the compasses, even L, whose mother thought she'd never seen one.

### Class 2

O and Knowledge: he knows that the sum of the angles in a triangle is 180°, he's known it for years. He sees no need to prove it, because it's a fact. "Do you want to prove that this hat is green?"

People are attached to their methods—there is the "copy a length by carrying it with a compass" school, and the "take another radius of the circle with the given length as radius" school. "You're just measuring." "Well, your line segment is attached to the original one."

There are now thirteen people in this section, but it works pretty well, because they are all willing to talk.

### Class 3

Great class. We were closing in on $\frac{\sqrt{5}-1}{2}$—. "618," says G, at regular intervals—and I'd say that everyone except P (and R who was absent) was involved all through. We hit the rational/irrational issue and faced it head on. L, of course, knew the decimal form for 1/7, and then we divided 7 into 1 (which was quite an event: several people had never seen a division algorithm—you either "just know" from times tables, or use a calculator) and saw that it repeats on a six-period and saw why it must repeat ("cool!"); so nonrepeaters can't be rationals—but maybe $\sqrt{5}$ repeats and we just don't know it. Decided to risk trying the proof by contradiction and we did it in full, to total satisfaction. "Weird kind of fraction!" repeated around the room. Even O was willing to say that "$\sqrt{5}$" was more useful notation than a decimal approximation—but G points out that "real scientists" use decimals, not fractions.

### Class 4

Another great class. Fascinating to find that last week's comfort with irrationals did not stretch to—or maybe prohibited—*dealing* with $\sqrt{5}-1$. "How can you subtract a Number from . . . an *irrational*?" Numbers should only do arithmetic with other Numbers. Well, what things are Numbers? Everyone agreed that positive integers were, but some doubt about negatives, because people had heard that minus times minus was plus and couldn't see why. Fractions unpleasant, but OK. Irrationals are exciting, and can describe lengths, but not clear that they are Numbers. Clear that we've got a very Pythagorean group here.

I raised the historical issue of doubt about 1 being a number (*a* nose, 2 eyes)—how long it took before 1 was admitted to citizenship, since numbers were thought of as collections of units, so the unit itself wasn't a number. Then of the difficulties with fractions—if 1 is the unbreakable (atomic) unit, 1/3 is a calculation, not a number. Or, as reflected in "numerator" and "denominator", *third* is a name even when 1 is a number. This is also a very verbal group, and their generous hearts allowed them to welcome $\sqrt{5}$. "As long," says L, "as it behaves like other numbers." "It behaves better than zero," says G.

Idle conversation about the meaning of Greek "atom", and the shock of an atom not being atomic, bore fruit later when G said a 7-gon was inconstructible—because he'd asked a scientist—ah, and what if you'd asked him about atoms? We slowly approach the idea that opinion is not identical to truth.

### Class 5

Lively, and full of unexpected stuff. Even the least mechanically adept produced recognizable 15-gons. To keep G from announcing one more time that he was finished, set him to turning his 15-gon into a 30-gon, hoping he'd see the limit is a circle.

### SECTION C (twelve-year-olds)

### Class 1

A: "The ideal triangle can't lie in the clouds because you could reach it in an airplane." R: "It would be so perfect it would evade you, always circling the globe opposite you." There's been a dramatic change in this class. Last term it was dominated by D's constant put-down of every question. Now T and K are set free to talk about ideas, U to ask for help with constructions. All sorts of loosely linked topics came up: the metric system vs. "normal": "God gave man a foot and an inch and a yard—" We were all amazed that it was ten minutes past the usual end of class.

### Class 2

Three new girls added to the eight prepubescent boys. They spent every second before class brushing their hair, and as much time as they dared reading magazines. Apparently one of them wanted to join the class but didn't want to come by herself, and her friends wanted to make it very clear that they were present only physically. Amazingly, the little guys never noticed—only K raised an eyebrow—when B had a screaming fight with her embarrassed mother.

F knows that the sum of the angles in a polygon goes up by 180° as the number of sides goes up. How does he know? It's in his math book, and it shows that it's true for up to 12 sides. Well, that beats "My Dad told me." Next time, deal with "true as far as 12"—and the difference between science proof and math proof.

### Class 3

A great class. Began shakily, with W veering out of control (a long anecdote about future people in a story shooting garbage into outer space) but F was jolly, attentive, cooperative; which somewhat eases my guilt at telling him off for antisocial behavior that he was probably totally unconscious of. At the end of class T, really concentrating, calculated the product of $\frac{\sqrt{5}+1}{2}$ and $\frac{\sqrt{5}-1}{2}$ with speed, accuracy, and delight in the solution. Wonderful side trips into Newtonian vs. Einsteinian physics, non-Euclidean geometry, the minimal ideal for axioms—the nicest sort of conversation with little boy touches, like A conversing with B by telling me what he would say to B.

### Class 4

A day of total madness. E's hexagon didn't close up, which convinced him that he could go around and around like an ammonite building on slightly-less-than 60° slices and so construct any n-gon. People pointed out that spiraling wouldn't create *any* closed figure, but he was sure that that problem was trivial. U waited until everybody was on to the next issue before saying he had no idea of what was going on—and did that *three times.* F assured me that even his relativity seminar was brainless today. His theory is that there is too much sunshine today for our northern brains to deal with.

### Class 5

V suddenly began to talk, and then, blushingly, Y. A silly, friendly conversation about having braces on your teeth. But they kept on talking when we started looking at the series of inscribed and circumscribed triangle, square, etc. My plan was to get to the idea of visually trapping a circle. F: "This looks like trigonometry!" "What's trigonometry?" OK—so we took that tack—defining terms, fitting angles and ratios of sides into our previously Euclidean world. Everyone was curious and involved, asking questions, making guesses.

    Why was this such a good class? I hadn't prepared it, so there was no place I was impatient to get to, and no program for people either to follow or rebel against. We were all enthusiastic and disorganized, so they were freer about making suggestions and kept the discussion focused. Their contributions were genuinely intended to be helpful, not attempts to "do well". It was a real experience of being mathematicians working on a problem, not teacher and students in a class.

*   *   *

Here are some notes from a fall course for seven- to nine-year-olds. If the question ever came up of what the topic was, the answer was simply: "Big Numbers".

### Class 1

Only five at the first class—four of them *extremely* talkative, relaxed, and joking. It's going to take a great deal of care in formulating questions to keep from descending into chaos. It was, for example, a mistake asking "What's the biggest number you know," expecting a trillion or a googolplex—and producing 45 seconds of 99999. . . . At the school W goes to they celebrate the hundredth day of school by having each child bring in 100 of something. W planned to take in his toy cars but lost track at 75. Another revelation: seven-year-olds talk with as much sophistication as eight-year-olds, but they have real trouble reading handwriting on the blackboard—implying I should print up handouts. But how can you print up in advance what hasn't happened yet?

### Class 2

How does one cure a migraine? Spend an hour with six yammering children. The Role of Adrenalin in Teaching. We started today with how to write big numbers in small spaces by using exponents—on our way to Archimedes and his myriads of myriads. Fell into a developmental cleft: N and W hadn't a clue. O and J got it instantly. It was right on target for E and S: they like structure applied to enormous size. They were both troubled by the idea of a finite universe (my image was of a crystal sphere slowly being filled with sand, like the bottom of an hour glass). S: "But where would you *be*?" E: "How can they say how many particles in the universe? How can they measure?" W has the solution: hire a scientist to count the grains.

T claimed that $3(x + 2)$ was the same as $3x + 6$, but C said they weren't. T said yes they were, it worked every time. A: "You haven't tried it for every number!" I put in my oar: "Remember when we figured out an easy way to do $5 \times 107$ as $5 \times 100 + 5 \times 7$?"

S: "But that wouldn't work for small numbers!"

### Class 3

Working on Kaprekar's constant.* Amazing concentration with which they went through seven successive steps to arrive at the cycle. N, S, and W now sit together in a triangle, one in the front row, two in the second, as far away as possible from J and O, who sit at the other end of the front row. E sits alone in the third row, but on the J and O side. This allows me to teach three separate sections—each knows when its eye is being caught for a sophistication-appropriate question. The seven-year-olds had never seen borrowing, so we did it in detail: theory, practice, chant. Nine-year-olds are good subtractors, and E's speed kept O from complaining of boredom. N (randomly as usual) announced that $\sqrt{1000}$ was 100—his

Kaprekar's constant: Weird!

*Kaprekar's constant: take any four-digit number (not all digits the same); arrange it twice, once from highest to lowest, once from lowest to highest. Subtract. Did you get 6174? If not, repeat the process with what you did get, and go on: you will inevitably arrive at 6174. which is divisible by 9.

215
$6174/9 = 686$

calculator said so. We disproved that quickly, then tried to find what $\sqrt{1000}$ *was*. $10 < \sqrt{1000} < 100$, then narrowing—quite reasonably, to $30 < \sqrt{1000} < 40$—and again impressive determination. J: "Try 33 1/3!" —which required totally new techniques. We got to $3\ 1/8 < \sqrt{10} < 3\ 1/4$ and left it for the future.

### Class 4

S happily calculated $2^{2^0}$, but could see no sense in $2^{2^k}$, because k was a letter, not a number—this despite his former happiness with letting x take on a range of values when playing function machines. So I said Fermat's "formula" was just another function machine, but he preferred to use k for his inputs. S was then happy, but I wasn't, sensing an abyss just glided over. Wrote "variable" on the board, and asked if anyone knew what that word meant. O: "It means different kinds of food. We're doing an experiment at school and the variable is what they eat." Children and language learning: they are programmed to pick up and use new words they have a rough sense of by context, but their using them by and large correctly is no demonstration of their knowing their actual meaning. A real issue in math, where there is a great deal of "naming of parts", easily confused with learning concepts.

We forget what a miasma of myths our certainties emerge from: error and hearsay maintained through endless generations of eight-year-olds. We had turned to function machines when concentration lapsed, and C's secret rule was: "times 2". S put in −1 and all hell broke loose. C calculated to himself, then announced: "You don't realize how bad this is, because you get two different answers. If you say $−1 \times 2$ you get −2, but if you say $2 \times −1$, you get +2, because when the minus is on the right, it reverses, because you're going backward."

"No," said T, "it's no problem at all, because $2 \times −1$ is $−1 + −1$, and that's 0."

"It isn't the sign that reverses," said A, "it's what you do to the numbers. When the minus is on the right, times turns into divide, and 1 over 2 is 1/2."

### Class 5

A mad run through what is now comfortably seen as Fermat's function machine. We'd got to $2^{2^5}$, i.e., $2^{32} + 1$, and doubled our way to 4,294,967,296 +1, decided we'd have to check all primes less than 65,536— and then found it *wasn't* a prime. O was outraged at Fermat's stupidity. J, the tinkerer, tried to persuade her that it was a very good guess, and that you have to make guesses and test them and expect that some of them will be wrong—but O was having none of it. What made her really mad was that she had put a lot of work into something that Fermat already knew was wrong: he shouldn't be wasting her time like that—she has enough to do learning true stuff. Nervously moved on to Mersenne primes, hoping she'd allow their provisional nature. As people calculated them I wrote up, by chance in columns,

$$2^1 - 1 = 1$$
$$2^2 - 1 = 3$$
$$2^3 - 1 = 7$$

E saw a pattern: "You just double and add one for the next output." Amazement and joy—no more need to raise 2 to outrageous powers. But why does it work? Well, $2(2^n - 1) + 1 = 2^{n+1} - 1$, but how extract that abstraction from an eight-year-old, whose interest is in *does* it work, not why.

## Class 6

O and J out of town, so could focus on the math-level of the young, and since N was there, the seesaw was very tipped in the seven-year-old direction. We got (reanimating, with difficulty, the calculation of $2^m \times 2^n$, a tool the older kids specialize in) powers of two worked out through $2^{20}$, and they would have gone on for the rest of the hour, had I not suggested relaxing by switching to powers of 10. I always forget how determined the young are to write out all the zeros (big numbers, like dinosaurs, as dangerous and powerful friends). $10^{20}$ was, I claimed, such an easy way to write a number whose name is never used and which is never written out—and they were hard at work writing a "complete" list of the powers of ten, their expansions, and their names. A different idea of control of the world through notation. Moral: in teaching ideas about math I'm very careful not to impose my understanding on their experimentation—but when it comes to wallowing in computation I tend to forget to repress my views.

## Class 10

Everything worked and all our simmering ideas came to a boil, so to celebrate we ended with an orgy of function machines. In high spirits, W put a canoe, two people, and a paddle into C's machine, and C (who takes everything seriously) said he needed to multiply human times human and didn't know what he'd get. "Human$^2$," said J. "Yes—but what *is* that?" No one had anything to say, until E offered: "Human divided by human would be 1, because anything divided by itself is 1."

"Of course!" said C—then: "No! I wanted to multiply!"

I said: "I think it's a complicated question. I just saw a book that said \$2/2 = \$1, but \$2/\$2 = 1." Everyone said "Of course."

I went on: "But I don't understand what \$2/\$2 means."

"Nobody knows what it means," said A, "but everyone knows the answer is still 1."

I should have asked them whether \$2 × \$2 was four square dollars.

*   *   *

From a journal about two spring semester courses dealing with solving equations.

Thirteen-year-olds (and two parents) were in Section A, fourteen- and fifteen-year-olds in B.

## SECTION A

### Class 1

The issue came up of whether infinity is completed or potential. N: "You just make one when you need one." Counterview: they exist in the mind of God (eternal, omniscient, and won't get filled up). N's father, P: "So all infinities are equal?" G (another father and Math Circle veteran): "That's another course."

J: "Are they called rational because they make sense?" I tried to explain ratios using "If you need two cups of water for one cup of rice ..." but this is an all-male class, had never cooked, and couldn't think of any other instance in which you might think in ratios.

What if you had 1/2 over 12/7? "That wouldn't be a rational, because it's not integer over integer." K: "It would be if you multiplied top and bottom by 2 and then by 7." This raised the whole question of *time*: Does that expression *become* rational, or was it rational all the time but just "invisible"?

We fell into the trap of calculation. People wanted to get "one more" decimal place in trapping a $< \sqrt{2} <$ b—"I'm nearly finished." It takes a person who dislikes arithmetic to say: "The method is clear and working it out is tedious." The nonrepeating issue came up: how would one ever know? "How many places would you have to calculate to be sure it didn't repeat?" K: "As many as there are counting numbers and we know we won't get there." J: "So we'll never know." We part in a combination of gloom and awe.

### Class 2

J comes in saying that what we have to do with $\sqrt{2}$ is to prove that it *is* a fraction, and not try to catch it between decimals. N, groaning: "The task would be unending." I wrote "$\sqrt{2}$ = a/b" on the board and said "All we have to do is find a and b"—and slid in the idea that the *real* a/b should be in lowest terms to keep us from getting an infinite collection of the same thing.

Instant demand that we "square both sides" and then that we "multiply by b²." Why? "It's much easier to see things when the equations are simple." But nobody saw anything: we stared at $2b^2 = a^2$, people tried putting in values for a and b, unsuccessfully. (Interesting point: nobody tried substitution for a/b, because they were convinced from last time that that was a mug's game—but once we had it in a different setting, that conviction faltered).

P: "Stop! The Pythagorean Theorem proves it's impossible!" His idea was that in the right triangle 1,1, $\sqrt{2}$ , a and b both equal 1, and since 2 ≠ 1, $2b^2$ couldn't possibly $= a^2$. I drew it on the blackboard and let it stand. N: "They're different variables, Dad." Impasse.

Me: "What if I floated the words 'even' and 'odd'?" Universal belief that the square of an odd is odd, of an even is even. "How do you know that?" "It's a fact." "What if someone didn't know that?" "He'd be stupid." I seized the opportunity for a disquisition on the difference between ignorance and stupidity, but it will recur—thirteen-year-olds equate intelligence and knowledge, so logically must equate stupidity and ignorance. The usual problem of correlation and causation.

"Could you prove an odd squared is an odd?" Everybody starts calling out odd numbers and their odd squares—but N says "Stop! We'd have to do N of them!" Great! Quickly they decide $2n + 1$ is odd, and "the square of that is $4n + 1$, and that's odd too." Too close to the end of class to take up that whole issue, but not a good idea to let it pass. Me: "OK, if n=3, then 2n +1 = 7; what do we get for its square?" "13." Horrified Chorus: "But that's wrong!" Chorus of those who were annoyed that examples don't prove: "You can just give an example!" Then the bell rang.

## Class 3

No meeting last week, but somehow the class has gelled. Everyone talked, not just the veterans. Started with the old problem of squaring $2n + 1$, as paralleled by squaring $7 = 2(3) +1$; A and M became responsible for arithmetic. Although K is convinced that it is a *fact* that $(a + b)^2 = a^2 + b^2$, he admits that it doesn't seem to be coming out right today. He's going to check on it.

The proof that no fraction equals $\sqrt{2}$ rolled up so quickly that I thought I'd better recap—no need. High energy and general participation has led to understanding.

But then, having successfully lined out $(x - 3)^2 = 4$ as $x^2 - 6x + 9 = 4$, and moved on to $x^2 - 6x + 5 = 0$—when I asked if, coming into the room and seeing that expression on the blackboard you could turn it back into "a square = a number", they were totally flummoxed, and went for $x^2 = 6x - 5$ and even $x(x - 6) = -5$, which, said D, was impossible, because the product of two numbers couldn't be negative. Sometimes you feel you are coaxing butterflies to march.

### Class 4

Moving from the raw material, what sorts of numbers there are, to putting them to work in functions, we got into translating the geometric idea "two points determine a line" into an algebraic statement about (a,b) and (c,d) => y = mx + k.

Thanks to giving directions, we got the map idea that a point is determined by how far up it is and how far over, and thanks to people's having been to New York, got a grid with streets across and avenues up and down, and even distances as either on foot ("add up all the ups and all the overs") or crow flies (Pythagoras)—but everything crashed at the transition from numbered points to abstract coordinates. We got a good definition of an unknown constant: "I have a birthday, you don't know it, but it's fixed—and you know *that*." So we could get a line determined by two fixed points, but couldn't get a way of describing *any* point on the line. Did get that "a line has to continue in the same way," or "at the same angle." Impasse—indicated by people repeating what they'd said before. Math as politics.

I decided to break the wrangle by talking about railroads and asking why they couldn't climb up steep slopes. Rapidly produced "If they wanted to get over the Rockies they'd have to build a trestle starting at the Mississippi," and the idea not of angle but of rise/run. Everybody had a steepness story, either of climbing or of skiing—general conversation, building the concept of slope into a physical context. And then K said: "If you go down hill do you call it negative slope?" and we segued right into abstraction. Another day when people wouldn't leave, even for juice and cookies—they wanted to figure out: if you have two lines, can they only be parallel or intersecting at one point? What about two lines lying entirely on top of one another? Do they become one line?

### Class 10

Great breakthrough! K insisted we solve a quadratic by formula—"the real way"—and not waste our time completing the square. He stated it and began plugging in, and D (the guy who began the year with "I know, because somebody told me") insisted that K explain: "Where does it come from? How does it work? What justifies it?" So we solved $5x^2 - 24x + 27$ across the four blackboards in parallel: (a) by factoring, (b) by completing the square—then we solved $ax^2 + bx + c = 0$ by completing the square to produce the formula, and then (c) everyone agreed to allow K to use the formula—and the class ended on a note of universal satisfaction. Nicely, people have different favorite methods; the formula hasn't conquered the homemade.

220

**SECTION B**

*Class 1*

Conversation on the pains and delights of symbolism, inspired by Bhramagupta's paragraph-long quadratic formula: how one yearns to tell him ours, vs. how annoyed one is at seeing some symbol one doesn't understand. "My sister's calculus book is full of long wiggles." One of the Russian girls: "It's like having things that are easier to say in Russian and things that are easier to say in English." Impressive that they already see that math is *expressed* in symbols, rather than *being* symbols.

I had on the board "$x - b/a = 0$", and then wrote $x = -b$. W: "How did you get $-b$?" Me: "Because I'm an idiot." Delighted laughter—at the implied safety of asking a question, implied acceptability of making mistakes, at the ridiculousness of a typo being equated with idiocy—a moment worth twenty repetitions of "collegial, not competitive." But joking isn't universally valuable: in the other class D and L, who have no sense of irony, would just say, "Teacher admits to being an idiot." A difficult balancing act.

A good example of the values of symbolism: S "couldn't understand" completing the square of $x^2 - 64x$ because R had said "1024" so quickly; but $x^2 - 2ax$ got through. How frequently slow arithmetic produces a sense of incompetence, and hence readiness to skip the whole thing. Does that mean there should be more emphasis on mental arithmetic in the younger grades, or more abstraction? Rote certainly leads to boredom—and uses up a lot of valuable time—but abstraction may not be compatible with brain maturation.

*Class 2*

The question arose of why the History of Math concentrates on different places at different times. Was nothing happening in other places? Did the Greeks just stop after Geometry? Where is math now? What did Islamic math have that nobody else had at the time? Answer: both Algebra and the idea of proof. The interesting phenomenon of a slightly different topic bringing new people into the discussion—and once people have started talking, they tend to continue. A good argument for widely read teachers.

*Class 3*

Discussion of whether math is *truth* or is culturally determined (arose in the context of clay tablet Plimpton 322). E is convinced that it is all created in the mind of man, and only technologically determined insofar as tasks suggest realms of invention. There is no external truth. B says

3 – 2 = 1 is a universal truth. E and M argue about imaginary numbers—E that they are no more imaginary than any other numbers, M that the mathematical properties of the other numbers are intuitive. T says that minus times minus is plus is just as artificially designed, so negative integers are just as imaginary as imaginaries. T had a lot of trouble making herself heard over P, E, and V. That's a real problem for girls in math: having to shout and hit, which they learn to stop doing by age ten (it doesn't happen as much in other subjects, because boys are less interested—or argumentative—in discussing literature or history).

### Class 4

Working on finding middle terms of Hippocrates' cube doubling, leading to Appollonius' conic section solution. Good questions about the value of stating hypotheses even if you couldn't prove them (leading to a digression about Fermat and Wiles). Then I asked: "If you see 'ratio', where do you go to look?"—expecting similar triangles—and A said "trig functions," and then W asked whether they were really functions: how, for example, does a calculator answer when you ask for sine 32°? (The great usefulness of giving children calculators lies in all the applications they don't know how to use, which really exercise their imaginations—they play around and try to make sense of what comes up). My taste is geometric, so instead of curves I drew triangles—getting 45°, 30°, 60°—and at $(\sqrt{2})/2$ coming out of the calculator as .7071, B raised the Greek objection to approximation.

Back to conic sections with cone and slicing planes, good arguments about parabola vs. hyperbola, limiting cases (circle just one of a set of ellipses), everybody popping confidently with contributions.

N: "I'm not particularly religious, but those numbers today could persuade you."

### Class 5

Moving toward the idea of a cubic formula—a machine into which one could plug data—and to the graphic representation of cubics, with al Tusi's idea of the local max of a cubic being based on the max of a quadratic at $x = -b/2a$. The class split instantly into algebraists vs. geometers, with each convinced that reality lay in only one of the two formulations.

### Class 10

Moving through the permutations of symbols to produce the cubic formula: an amazing universal balk at change of variable. Everyone agrees that it would be *useful* to replace x with y – b/3, but it seems immoral (I'm glad I didn't follow the convention that replaces x with x – 3: that

*(margin note: Here's al Tusi!)*

would have turned them off math forever!). Very important in planning a course, a class, to predict the places where people will hit the "do bats eat cats" barrier, where words (or symbols) have stopped conveying meaning and just turn into sounds. Especially because some students are so good at memorizing nonsense syllables that you might not notice that they don't know what they're saying. Another argument for conversation.

*　*　*

I hope this journal gives some sense of the exhilaration of teaching in The Math Circle style. It requires preparation so thorough that it's completely internalized, being able to turn on a dime, sharing the excitement of the chase, thinking on several levels, and from different standpoints. Where did we ever get the idea that teaching should be predictable and uneventful, with challenge and excitement saved for weekends and vacations? While colleagues are counting the time ("Only thirty-two more years!") to retirement, after decades at the chalkface, I still enter a classroom with the same anticipation I felt on my first day of teaching.

## What's the Way Forward?

The Math Circle will spring up, as it has in our neck of the woods, when the demand of enough dissatisfied parents finds a supply of interested teachers. But this leaves too much to chance, and the luck of finding energetic enthusiasts, for it to spread very widely. We can think of several other ways for it to take hold.

Loose networks, like those of home-schoolers, might form more or less spontaneously, drawing on people whose different agendas coincided perhaps in no more than the desire to have math developed collegially, and in depth. As you saw from the Competition section in chapter 8, there are already associations of competitive math circles, with meets and prizes. These would be unlikely to morph into our approach, but families and teachers who are engaged in them, because they are the only option at present, might welcome the chance to join a more collegial group. A problem with spontaneous groups, however, is their dependence on individuals: doctrinal differences might well swamp them, and another Movement would have run its course.

A tighter network, as of after-school Math Clubs meeting on school grounds, would be a second possibility. Were these funded and overseen by some central organization, like an existing mathematical body, some sort of cohesion might follow. Drawbacks include all those of centralized bureaucracies, to which the freedom of mathematics would be especially vulnerable. In addition, by being in the after-school slot, competing for students' time with sports and activities, their reach would likely be limited to the few already enthusiastic about math. Better still would be an informal federation of Math Circles (sharing topics, successes, and failures via the Internet—and perhaps visits in person), each run in its own way, with courses chosen by and suited to those in it (math pure or applied, leading to or from the sciences). For the sake of that freedom so central to mathematics, keep administration and hierarchy to a minimum. Each ship, then, on its own bottom—but all flying the flags that signal collegiality rather than competition, serious content, and discovery rather than instruction.

The best way, certainly, would be for The Math Circle's approach to replace that in regular school classes—and this will surely happen one day. The question is how to move that day closer to this. The steps are straightforward. For generations, school math curricula, from twelfth grade down to first, have been largely determined by college entrance requirements; and here The Math Circle style of teaching works far better than the present curricula, with their petrified remnants of business math and practical math—poorly combined with math for fun and math worksheets to free the teacher for other tasks.

We've seen our students consistently master what's thought to be very advanced material, because they wrestle and think their way through it, propelled by the fun of the chase. A course begins slowly, as people mull the problem over, but then speeds up with the excitement of discovery, and no time is wasted on review because it is their intellectual property. Nothing is memorized, because it is understood: in math, the need to memorize testifies to not understanding. A model to keep in mind is the old shop class: you don't have to memorize whether a saw, pliers, or a hammer is most conveniently used for driving in a nail. The level aimed for isn't cabinet-making, but it isn't wood-butchery either: solid carpentry lets you advance.

As always, we come down to the problem of developing teachers able to run such courses. Math has traditionally been the least favorite subject of elementary school teachers; few have studied more math than what they had in high school, and even fewer have a real interest in it. In order to teach a flexible and inventive math course, they would have to master a lot more math, not just a new methodology. The job would have to be made attractive, both in pay and esteem from colleagues and the community—and these political and social issues rest ultimately on the perceived need for a mathematically literate population (a need that grows as automata take over less skilled jobs). Given the body of people eager to teach, how prepare them? By using The Math Circle approach in teachers' colleges. We've now narrowed the problem down to the much smaller population of those able to teach the teachers. *Yes.*

Perhaps—as in countries where literacy was spread by insisting that whoever could read should teach two more—those who will teach the teachers will come from the ranks of Math Circle teachers and students. Three of our graduates have already taught courses, and our colleague Jim Tanton now runs just such a training program for future teachers. Perhaps this book will inspire others, in who knows how many different ways: mathematics is freedom.

Whenever we've given a demonstration class, conversation afterward in the teachers' lounge followed the same pattern: surprise at the number of students who got involved ("Sally never says a word"), amazement at the sophistication their students could handle ("They can't reliably add 7 and 3—and here they were figuring out the sum of the first hundred integers in no time!"), envy of the fun we were having—and then, always: "But of course I couldn't do anything like that; we can barely get through the syllabus as it is." *Chomsky is a control freak.*

The Math Circle has, indeed, the luxury of an after-school program's freedom from responsibility, but we too have taught in K–12 schools and universities, and know that you are never free from the syllabus, the exam, the necessity of preparing for the next course (we never are, but always to be, blessed with boundless exploration). Is it really impossible to harness the energy of The Math Circle to the cart of meeting requirements?

Not at all. The greatest innovation takes the least time. Instead of stating a rule and then assigning "exercises" which use it, start with the problem and work as a group to discover the rule: not "A = bh/2, what is A if b = 3 and h = 5?" but "Here's a triangle; how could we find its area?"—so turning conscripts into colleagues. It may take ten minutes to come up with the answer, but no one will ever "forget the formula" when they can re-create it as they need it. That works as well for AP Calculus as for fourth-grade Shapes and their Properties—a technique one devised oneself won't revert to symbol-pushing after a week. Curiosity, and pride in workmanship, will have been awoken.

With this approach, the term begins with your class running behind the syllabus, but it catches up about midyear—and has lots of spare time in the weeks usually devoted to review before the exam. But what if you have to pass a checkpoint on time every week, as in a course with many sections and common tests? Then you can divide the class period into meat-and-potatoes and dessert: first a quick and active discussion of the day's topic, still letting the students discover, but directing the search a bit more—and then as a reward, a completely free discussion of some ongoing unrequired topic (the Web is a good source). The promised reward focuses attention, and you have Math Circle minutes, if not hours.

Another great time-saver is intellectual honesty: "Why do we have to learn this technique?" is a fair question, and should be answered to the best of one's ability—how its use helps, how they could have developed it themselves—and then the revolt, which was unspoken—"They're just making us memorize nonsense"—subsides, and the next topic will be greeted with an open mind.

There is one more serious problem that classroom teachers face and that no approach can entirely solve: classes of over twenty-five inevitably decline into lectures, with a few active students and a large passive back of the room. What we are learning about working memory shows that for many people, being told something registers far less well than doing it oneself, so lecturing is about the least effective form of teaching. If economics dictates that large classes are more important than effective classes, call on the good will of local college math departments and their students, on parents, on fellow teachers with a free period, or on some of your students—and split the class at least part of the time into small, problem-solving groups.

And what can parents do? Agitate for smaller classes—and when you're not engaging with the local school board, encourage your child's interest in math. You don't have to know much yourself to puzzle together over whether there are more counting numbers than even numbers, or how to find the balance-point of a triangle, or what the odds are of getting a royal flush: such conversations often draw the generations together. Steal a few minutes on the Web to find good questions, and even if they come with answers, try letting your child figure it out, with the fewest nudges: a fine exercise in patience and good humor. It's also fun to play Fermi Estimates (named for their inventor, the Nobel laureate, Enrico Fermi): how could you make an enlightened guess at the number of piano tuners in the greater Cleveland area? How many slates are there on the roofs of your town? How many stars can we see? How many hairs on the average head, or words on a newspaper page? How long would it take to drive from here to Maracaibo? These games sharpen the number sense and once again, using the Web, help explore the world.

The Fermi solution

Let's think—students, parents, and teachers alike—more broadly and boldly. Why confine this approach through imagination and thoughtfulness to mathematics alone? Why shouldn't history, literature, languages, and the sciences be subjects of collegial inquiry from the start—savoring their phenomena and making them accessible through reason rather than memory, engrossing ourselves in their substance rather than diminishing it to multiple-choice scores?

226

## To Take with You

We have an inborn urge to know, to make, and to enjoy the world around us, that neither fatigue nor fear can long suppress. It shows up as wonder that turns to awe; as a knack for engineering; and as what we've called the architectural instinct. These all, as they draw us on, reveal a labyrinth of shape and number that we move through. Daedalus made his wings out of feathers and wax; we make ours out of observation and reflection—and so rise out of the labyrinth, and see the whole world spread below us.

*Ah! Here's their guiding metaphor and the title of the book.*

# Appendix

## Thoughts of a Young Teacher

Sam Lichtenstein was a student in The Math Circle every semester from the age of ten. When he was seventeen and a high school junior, he struck us as a potentially wonderful leader—and so indeed he turned out to be. Here are some thoughts of his, written in December 2004, his second year of working with us.

\* \* \*

In elementary school, my friends and I would compete to see who could do our quizzes on multiplication and division the fastest. In retrospect, this might have been a bit of a burden on my teachers—how were they to teach math when some of the students were quite literally racing ahead? Subsequently, my teachers arrived at a solution: give me and the other mathy kids puzzles, puzzles, and more puzzles. The best, and most prominent, of these puzzles were "logic grids". An example of the type of question these asked would be to match up a group of kids with their hat color—say, figure out who among Billy, Yasmine, and Greg has a Blue hat, who has a Yellow one, and who has a Green one. The only tools we were given would be a series of statements, such as "No kid has a hat whose color begins with the same letter as their name." And with these statements, we were to figure out everything, using logic.

The reason they were called "logic grids" was that there was a set strategy for such puzzles: write a table with the kids' names on one axis and the colors on the other axis. Then in the appropriate box of the grid, place an X when a situation (e.g., Billy's hat being Blue) is logically inconsistent with the given statements. Using a process of elimination, the solutions were generally straightforward. Now that I think about it, it seems likely that some of the more difficult puzzles required the use of some rather sophisticated, yet still intuitively obvious, concepts, such as the Pigeonhole Principle. But at the time, it all fell under the heading of "logic", and it was a good way to spend a day of school. Unfortunately, after the hundredth or so puzzle, they started to seem repetitive.

But then I discovered The Math Circle.

I don't know how my parents found out about it, but they did, and I started going. I must have had a great time, because I put up with conflicts with soccer practice, Hebrew school, etc., in order to attend class at Harvard on Wednesday evenings. And in fact, I *know* I had a great time, because I remember some of the fantastic things we were learning. Cantorian Set Theory stands out, of course; so do ruler-and-compass constructions and complex numbers. All of it appealed to me and got me *excited* about math—which was, it turned out, so much more than logic grids and multiplication tables. It was Infinity, it was Games like constructing things with a ruler and compass, it was Fantasy like "imaginary" numbers! In short, it was *cool*. So I kept going, and I've never stopped.

Meanwhile, I was learning some okay stuff in math at school, but nothing really exciting until sixth grade when a particularly good teacher gave me extra algebra to do, setting me on a path to take more advanced classes in middle school and high school. Basically, I got lucky. The difference between a good teacher and a bad one was what defined my public school math career. Looking back on it, it was a near thing! And this is why programs like The Math Circle are potentially so valuable: they have the power to transform math education from a gamble into a delight.

Bob and Ellen Kaplan and Jim Tanton got me hooked at the Wednesday classes. The style was pretty much perfect for me; as I saw fit I could participate or observe, believe or question, conjecture, theorize, or experiment on my own. Most importantly, I was tackling problems and ideas that were fascinating and expertly presented. At some point, I switched up to Sundays, and the classes got cooler—if that's even possible. Knot theory from Bob in addition to Jim's regular topological tidbits, Babylonian mathematics from Ellen (the sophistication of which I am really only coming to appreciate now!), a proof of Gauss's Fundamental Theorem of Algebra using winding numbers that was, if not totally understandable, at least *accessible* to a middle schooler. It was awesome, and that's why I keep going.

Then, in 2004, Bob and Ellen invited me to teach a class for the younger students. I almost literally jumped at the opportunity—I was going back to the Wednesday night classes I had loved so much back in elementary school. Boy, was I in for a surprise!

I had done a summer program (PROMYS) at Boston University in 2003, and among the topics covered were Simple Continued Fractions. I liked the fact that the continued fraction

$$\cfrac{1}{1+\cfrac{1}{1+\cfrac{1}{1+1\dots}}} = \frac{1+\sqrt{5}}{2}$$

*And the left side of this I still don't understand.*

is the Golden Mean you see in Fibonacci numbers, rectangles, pentagons, phyllotaxis, and various other miraculous mathematical phenomena. So I planned my class starting with this fact, and it turned out my students liked it too! (Ah, success!)

As the semester progressed, our investigation of simple continued fractions took us far afield. This was partly due to my own inexperience as a teacher: SCFs are (in my opinion) a pretty difficult number theory topic, and I neglected to take into account the tools of number theory. These are of two sorts. The most important are observation, experimentation, and curiosity. This was fine; the kids had all these things in spades. But the second most important tool is algebra, and this was something the kids were a little less comfortable with.

I tried to think back to my own experiences as a Wednesday nighter. Obviously, I *couldn't* have known algebra then. But we covered topics that can only really be considered algebraic (for example, the algebraic properties of Cantor's transfinite ordinals, which lack commutativity). How did Bob and Ellen and Jim swing that? It seemed that they must have been tricky about it, sneaking the abstractions and generalizations up on the students, after we were so comfortable with the concrete examples and knew the rules of their behavior so intimately, that the abstraction—whether or not we comprehended the algebraic symbols—was merely an expression of our gut feelings.

Well, this is easier said than done, and the Kaplans are a tough act to follow. (After all, they had been doing this a tad longer than I had.) But I did my best, and we made some pretty decent progress on SCFs, although we fell short of my admittedly overambitious goal of proving the theorem of Lagrange: that any quadratic surd has an eventually periodic continued fraction. No matter, the kids were moving on. Someone mentioned the classic puzzle of connecting three houses to three utilities without crossing the pipes. And thanking the gods for dropping a topic on me just when it seemed I was empty-handed, halfway through the course, I began introducing the students to graph theory.

*Oh really?*

After some work, we solved the Utilities puzzle, and solved it completely. (I'll leave it as an Exercise to the Reader, because it is an excellent puzzle.) But where to go from there? I knew and the students knew that Math Circle is not about presenting a subject—such as graph theory—in a condensed format, but rather about *playing* with the subject, fooling around. So that's what we did; we played games: first games with

Conway's Nim-games

graphs, and then soon all kinds of games. Before I knew it, we were investigating Conway's Nim-games (games of full information, which is to say, no element of chance) and their associated strategies. And thus the class grew into its own. We were really into the swing of things, when, sadly, the semester ended.

I took a deep breath. That was *way* harder than I thought it would be! (I'll always look at Bob and Ellen and Jim and the others with a little more awe now, knowing that they taught several Math Circle classes each week, and were also teachers—and prophets of The Math Circle—in "real life". Sheer madness!) But I knew I had to do it again.

This semester I've been teaching about modular arithmetic, repeating decimals, and cryptography. With a little more experience, I've been a little more comfortable. But I've also made a big breakthrough in my teaching recently. When I first started, I planned my lessons extensively—everything from topics I wanted to cover to order and manner of presentation to activities to keep the kids engaged. Recently, though, I've started taking risks, reading up on a topic, exploring a few interesting pathways, and going into the week's class "blind". And strangely enough, The key often these have been the best classes. The key is *lack* of planning, having fun, and just fooling around along with the kids.

I think this secret I've stumbled upon is the Secret of the Kaplans, the secret that all the best Math Circle teachers have known—Jim and the rest. And the explanation is that Math Circle is all about fluidity, the kind of fluidity of ideas that only kids are really capable of. So teaching is not about "lowering" the level of the material to meet the kids, but rather about looking at it from the same angle they are.

I must have known this secret deep down when I was a Wednesday nighter myself, but forgotten it. Why might this be so? After all, I've been attending Sunday classes all through high school. Perhaps such forgetfulness is a symptom of "mathematical maturity": as we learn algebra more thoroughly, codify the rules of the game, and climb the ladder to its "ultimate achievement" (calculus class: from a standard high school perspective, the End of Math), we become set in our thinking. In school, these things we're manipulating (variables, operations, integrals, derivatives) become game pieces, perhaps something like Mancala stones, which we move from place to place in certain ways. At the next level, we abstract the stones and treat their behavior as stemming from their geological properties—we examine algebraic structures like groups and fields; we make rigorous such intuitive notions as limits and infinity.

Teaching Math Circle, I've been relearning how to just look at the stones, play with them, marvel at them. And boy am I glad!

Hardening of the categories

29 June 2007

# A Note on Our Pronouns

People are put off math by its confusing symbols. But are we always certain what common pronouns mean? In this book "I" has indifferently stood for either of us (unless context makes clear who is speaking). "We" usually means both of us, but again, context decided; we used it at times not in a royal but universal sense—our species, our kind, abbreviated to a common point of view. "We" never means "them", since the collegial spirit of The Math Circle keeps its participants on a par.

English needs a gender-neutral pronoun warmer than "one", less clumsy than "him/her", and less eccentric than the once-proposed "e". Waiting for the right shim to level our language, we've said "you" more often than not, and otherwise wandered along, scattering "he" and "she" and "his" and "hers" throughout the text, to give it the festive appearance of a spring lawn.

# Index

abacus, 120

abbreviations, 51. *See also* notation; symbols

abstraction: abstract algebra, 132, 155, 168; and algorithms, 128; and the architectural instinct, 115; and arithmetic, 10, 98, 102–3; Bronowski on, 19; and concision, 99, 160; incorporation of, 102; and intuition, 181, 183; and the language of math, 109; and maps, 182; and the Math Circle approach, 205, 217; and numbers, 101–2; and the origin of math, 126; and perspective, 189; and problem selection, 180; and the recursive character of math, 122–23, 149; and simple continued fractions, 231; and slopes, 220; and symbols, 81, 86, 88; and teaching style, 175

accelerated learning, 160–61

accreditation, 133

accuracy, 52. *See also* precision

addition: and abstraction, 98; counting numbers, 166; and fractions, 2, 8, 11, 102, 151–53, 166; and intuition, 184

administrative issues, 223–24

adolescence, 168

adult students, 166–67

adversity, 139

age of students: adolescence, 168; adult students, 166–67; and attention span, 12; and calculation, 217; and class composition, 166; and effects of school, 6–7; and enthusiasm, 206; and equations, 218; and flexibility of thought, 179; and gender issues, 168; and induction, 184; and the Math Circle approach, 1–4, 200–204, 209–23; and the myth of talent, 20–21; and obsession with numbers, 6, 11

algebra: and abstraction, 102–3; "algebraic amnesia," 110; and coordinate geometry, 66; and cubic formula, 222; Gauss's Fundamental Theorem, 230; and geometry, 126; and intuition, 135, 184; and lines, 220; and the Math Circle approach, 221; and math curricula, 132; and methods vs. approaches, 39; and problem selection, 176–

77; and the quadratic formula, 40–41; and reduction, 59; and symbols, 82, 88; and "The Faustian Offer," 112; and visualization, 150

algorithms: and abstraction, 128; and atomizing, 40; and cookbook math, 128; and the Math Circle approach, 212; and memorization, 134; and rote learning, 117–18

alienation, 103–6

al-Khwarizmi, Mohammad ibn Musa, 82–83

all-Siberian Math Olympiad, 142–43

altitudes, 53–54, 203

al-Tusi, Nasir al-Din, 222

ambiguity in math, 48–49. *See also* abstraction

ambition, 169. *See also* ego

ambivalence to math, 110–15

analogy, 17, 40, 66–70, 75, 103, 163

analysis/synthesis, 40

Analytical Engine, 52

ancient Greece, 120, 123, 126–27, 147, 196n, 221

Anderson, Poul, 81

angle bisectors, 53, 180, 202–3

Appolonius' conic section, 222

approaches vs. methods, 38–40, 156–57

*a priori* nature of math, 9, 164

aptitude for math, 15–23, 113, 193–94

Archimedes, 166, 215

architectural instinct: and abstraction, 115; and camaraderie, 175; and "cookbook math," 131; and the culture of mathematics, 105; and curiosity, 227; described, 5–6, 76–78; and ego, 167; fostering, 161–62; and intuition, 97, 183; and the language of math, 20, 109; and play, 159; and the *a priori* nature of math, 164; and problem selection, 178; and the relevance of math, 113, 115; and symbols, 88; and teaching, 117

Aristotle, 91, 94, 126

arithmetic. *See also specific operations*: and abstraction, 10, 98, 102–3; and age, 166; and ancient cultures, 120; and common

# Index

# Index

# Index

# Index

# Index

Borders Books
Princeton, NJ
Sun 21 Jan 2007
 $18.75 with 25% coupon
 + 1.31 tax